THE MIDLAND COUNTIES RAILWAY

CW01497566

Dedicated to the memory of
Charles R. Clinker
Kathleen M. Stephens
and
Thomas J. Stephens

First published 1989
by the Railway & Canal Historical Society
on the occasion of the 150th anniversary
of the opening of the first section of the railway
between Nottingham and Derby

THE
MIDLAND COUNTIES
RAILWAY

Compiled by
the Derby Railway History Research Group

Tutor: C. R. Clinker
S. L. Bragg, R. H. Fullager, J. B. Sheldon, K. M. Stephens
T. J. Stephens, P. S. Stevenson, S. Turner

Edited by P. S. Stevenson

RAILWAY AND CANAL HISTORICAL SOCIETY

First published 1989 by the Railway & Canal Historical Society

Registered Office: Fron Fawnog, Hafod Road, Gwernymynydd, Mold,
Clwyd CH7 5JS

British Library Cataloguing in Publication Data
The Midland Counties Railway
1. England. Railway services. Midland Counties Railway
history
I. Derby Railway History Research Group
385'. 0942

ISBN 0-901461-11-3

Design Malcolm Preskett
Typeset and printed in England by Hobbs the Printers
of Southampton

FRONT COVER
Marshalling a train on the Midland Counties Railway below
Nottingham Castle (see illustration no. 30)

BACK COVER
An excursion ticket issued by the Midland Counties on 22
June 1842 (see illustration no. 33)

Contents

Illustrations, Maps and Plans

Preface

As built, the Midland Counties Railway linked Nottingham and Derby with Loughborough, Leicester, and the London & Birmingham Railway at Rugby, so providing ready access to the capital. At Derby it shared station facilities with the North Midland and Birmingham & Derby Junction Railways, becoming amalgamated with these concerns to form the Midland Railway Company after an independent operating existence lasting only five years.

These five years, between May 1839 and May 1844, were however momentous in the history of the English railway system, seeing the foundation of an interconnected network of trunk routes and the evolution of soundly based management techniques. The gestation period of the Midland Counties line, spanning the years back to 1824, were even more remarkable, seeing the bringing to perfection of the steam locomotive as a sound mechanical device capable of transforming communications previously provided by a combination of services on indifferent roads and the inland waterways.

The history of the promotion, building and operation of this railway, which straddled such a critical period, is of sufficient importance fully to justify the decision of the Derby Railway History Research Group to undertake its compilation, no separate work having previously been devoted entirely to the subject.[1]

Coincident with the formation of the Railway & Canal Historical Society, during the latter months of 1954, Charles Clinker, elected first President, was lecturing upon railway history for the Universities of Birmingham and Nottingham. Courses in Warwick and Birmingham were quickly followed by those in Derby and Nottingham during the years 1954–56. It was as a direct result of the close friendships developed within the Derby class, in particular, that the East Midlands Group of the Society came to be founded in 1957. Reconstituted a few years later, the Group has made continued progress to the present time, so facilitating the eventual appearance of this manuscript.

On 30 June 1956 the members of the Derby district Workers' Educational Association class were informed that Nottingham University had agreed to give them full support as a tutorial research group with a view to investigating the history of the Midland Counties Railway. Newspapers and other contemporary journals were to be combed through in the Prince Charles room at the Art Gallery and the information abstracted put together with extracts taken from the minute books and other papers in the British Transport Historical Records collection. It was even envisaged that the results might be published by the University as the first in a new series of booklets on local history topics.

After abstraction of the sources referred to, members of the group undertook to put their material into a narrative form suitable for publication. Not until 1967, however, were three of these chapters first brought together into a collected typescript version for circulation amongst the members of the group and progress was then halted by the deaths of some of the key figures. The forthcoming celebration of the 150th anniversary of the opening of the first section of this railway, at the end of May 1839, to be organised by the East Midlands Group of the Society and others, seems a suitable occasion to launch this previously unpublished manuscript, now thoroughly revised by one of the original compilers. The generous assistance of John Gough, editor of the Society's *Journal,* in this final process is gratefully acknowledged, together with that provided over the years by the archivists, librarians, and staffs of the various institutions referred to in the notes. Publication of the text of the 1832 prospectus, together with the reports of William Jessop and Joseph Glynn, has been made possible through the ready loan of original material by Leslie Hales.

Assistance with the provision of illustrations is gratefully acknowledged to Sheila Cook and Dorothy Ritchie of the Local History Department, Nottinghamshire County Libraries, Angel Row, Nottingham, to John Cockroft of the Industrial Museums Service, Wollaton Hall, and Suella Postles of the Brewhouse Yard Museum, both parts of the Leisure Services Department, Nottingham City Council, to Adrian Henstock, Nottinghamshire County Archivist, Keith Reedman of Long Eaton, and Glyn Waite of the Transport Ticket Society.

Introduction

Once a vital part of George Hudson's empire, linking York and Leeds with Euston via Normanton, Derby, Leicester, and Rugby, and today forming over most of its original course a vital part of the Inter-City network, the Midland Counties Railway evolved from some of the earliest schemes for English trunk railways.

Nottingham had known primitive wooden railways as early as 1604,[1] when Huntingdon Beaumont of Coleorton in Leicestershire laid down his pioneer installation to bring Sir Francis Willoughby's coal from the Strelley pits to Wollaton Lane End.[2] In 1764 William Brown, a coal viewer from Newcastle upon Tyne, surveyed an unbuilt line to bring coal into Nottingham from the colliery at Nuthall. But it was not until 1837 that Thomas North introduced the steam locomotive to the area on his railway from the pits at Babbington, near Awsworth, to the landsale wharf at Cinderhill.[3] In the meantime a network of canals had been built to carry coal from the Erewash Valley onto the river Trent and into the county town, using railways as short feeder lines from the collieries.

It was at the behest of the Loughborough merchants, actively pursuing the canalisation of the river Soar from their town to its confluence with the Trent opposite Sawley in Derbyshire, that the Erewash Valley coalmasters were persuaded into joining forces to promote a canal from the river Trent at Sawley northwards through their valley to reach the collieries above Ilkeston. The entire line of waterway between Loughborough and Langley Mill was brought into use before the end of 1779 and became perhaps, in its heyday, the most rewarding investment of its kind. On the Erewash Canal the first dividend of $2\frac{1}{2}$ per cent was paid for the year ending 6 April 1783. By 1787 dividends had risen to 20 per cent and in 1794 they reached as much as 30 per cent. Although they then fell back for a while, by 1810 they were once more over 30 per cent, rising to 51.3 per cent on average during the period 1815–17 and finally peaking at 74 per cent in 1826. All this time, of course, the shares were trading at very high premiums. The Loughborough Navigation, which had cost comparatively little to construct, paid its first dividend of 5 per cent at the end of 1780 and, ten years later, had upped this to 20 per cent. The price of the shares and the dividends upon them rose to fantastic heights. From 20 per cent in 1790 dividends had doubled by 1797–99, doubled again by 1803–05 and, at their peak of around 154 per cent in 1827–29, just failed to double once more. By the time that the Midland Counties Railway was opened through Loughborough, in 1840, they were still running above 100 per cent.[4]

The potential for profits of this nature attracted other investors to promote the construction of further waterways, opening up more of the coalfield and facilitating the establishment of ironworks and other manufactories. From Langley Mill northwards to Codnor Park and Pinxton, and also through the Butterley Tunnel and Ambergate, the Cromford Canal opened up better access and brought coal to Sir Richard Arkwright's mills at Cromford in 1794. It also brought together the powerful engineering team of William Jessop and Benjamin Outram, who linked with the financiers Francis Beresford and John Wright to found the coal and iron trading enterprise at Butterley, which was to make such a significant contribution to future communication throughout the country. Established as Benjamin Outram & Co in 1790, the name of the company was eventually changed to *The Butterley Company* in 1807.[5]

Construction of the Cromford Canal was closely followed by that of the Nottingham Canal, designed to shorten the journey for coals carried between Langley Mill and Nottingham whilst also assisting to open up even further that part of the coalfield lying west of the town, from which for so many years it had previously drawn its staple supplies. The Grantham Canal spread these benefits eastwards while to the south of Langley Mill the making of the Nutbrook Canal assisted in the development of the collieries at Shipley and West Hallam and the establishment of ironworks, first at Dale Abbey and afterwards at Stanton and West Hallam; those along the Erewash Canal at Bennerley, Trowell and Hallam Fields were not established until the arrival of railways in the valley. The Derby Canal provided a useful connection between the Erewash and Trent & Mersey Canals independently of the river Trent.

South of Loughborough, the Leicester and Melton Mowbray Navigations continued the canalisation of the rivers Soar and Wreak and spawned such extensions as the Oakham and the Leicestershire & Northamptonshire Union Canals. By means of the latter and other extensions even farther south the Erewash Valley collieries were able to market their products ever more widely, the first attempt to open up the west Leicestershire coalfield having totally failed. The Charnwood Forest branch of the Leicester Navigation had been an integral part of that undertaking, coming into sporadic use from 1794 onwards. Blackbrook Reservoir was not finished and

filled for another three years and in 1799 it burst with disastrous results. Although attempts were made to restore the canal it was little used and, for the time being, the Leicestershire coalfield slumbered to await the coming of the railway.[6]

The problems with the Charnwood Forest Canal stemmed partially from the hybrid nature of its construction. To overcome the great difference in height of the ground between Loughborough and Nanpanton, William Jessop had laid down an iron edge railway, not the first of its kind but perhaps the first to use strengthened fish-bellied rails. By doing so he ensured that there would be no loss of lockage water from the canal itself. From Thringstone and Barrow Hill basins, at the western ends of the canal, other railways had been laid down to the collieries and limeworks. In the Erewash Valley coalfield most of the connections between pits and canals had been made using iron plate-rails, of which Benjamin Outram was the champion, having improved considerably upon the earlier ideas of John Curr and John Butler as to the soundness of their construction. Some of the earliest had been formed of wooden rails but the only significant departure from Outram's model lay in the provision of fish-bellied iron edge rails on the Mansfield & Pinxton Railway, constructed by William Jessop's son Josias in 1817–19.[7] The promotion of such an extensive canal feeder line as a separately constituted joint stock undertaking, whilst not the first railway to be authorised by Parliament, marked a significant watershed in the development of communications, both locally and nationally.

The year 1824 actually witnessed the rise and development of what W.W. Tomlinson referred to as the first *railway mania*.[8] For two years the Stockton & Darlington Railway Company, steadily carrying out its own plans, had he said been making a great experiment for the rest of the country, and men of discernment had recognised its practical import and come forward with similar projects. These started into notice in the public newspapers almost weekly, many of them designed to intersect the country to its fullest extent, over hills and valleys in all directions. It was estimated that there were twenty new railway schemes in agitation, representing a capital of £13,950,000, at the beginning of 1825.[9] A *grand trunk railway* between London and Edinburgh had been suggested several months before this, for the conveyance of goods and passengers by means of locomotive and stationary engines, the proposed line being taken past Bedford to Leeds, leaving a little on the west the towns of Northampton, Leicester, Loughborough, Nottingham, Mansfield, Chesterfield, Sheffield, and Barnsley, and on the east Huntingdon, Stamford, Worksop, Doncaster, and York, and, in its course northward of Leeds, passing at nearly equal distances between Carlisle and Newcastle.[10] The nation was said to be *railway mad* and "unquestionably," wrote Edward Baines in the *Leeds Mercury* of 24 December 1824, "the rage of speculation has taken so decided a turn in this direction, as to present several symptoms of the popular delusion which sometimes arises out of a strong and general excitement of the most active passions of human nature."

The next most ambitious scheme in conception was, perhaps, the London Northern Railroad Company, launched on 13 December 1824, with a capital of £2,500,000, "to connect Birmingham, Derby, Nottingham, Hull, and Manchester with each other, with the parts adjacent and with the metropolis."[11] It was with this scheme, suspended during the financial reverse at the end of 1825, that the idea for a railway between the collieries at Pinxton and Leicester may be said to have originated. In its first public pronouncements the Midland Counties Railway made it clear that it had adapted the plans made by Josias Jessop for the London Northern Railroad in 1824.

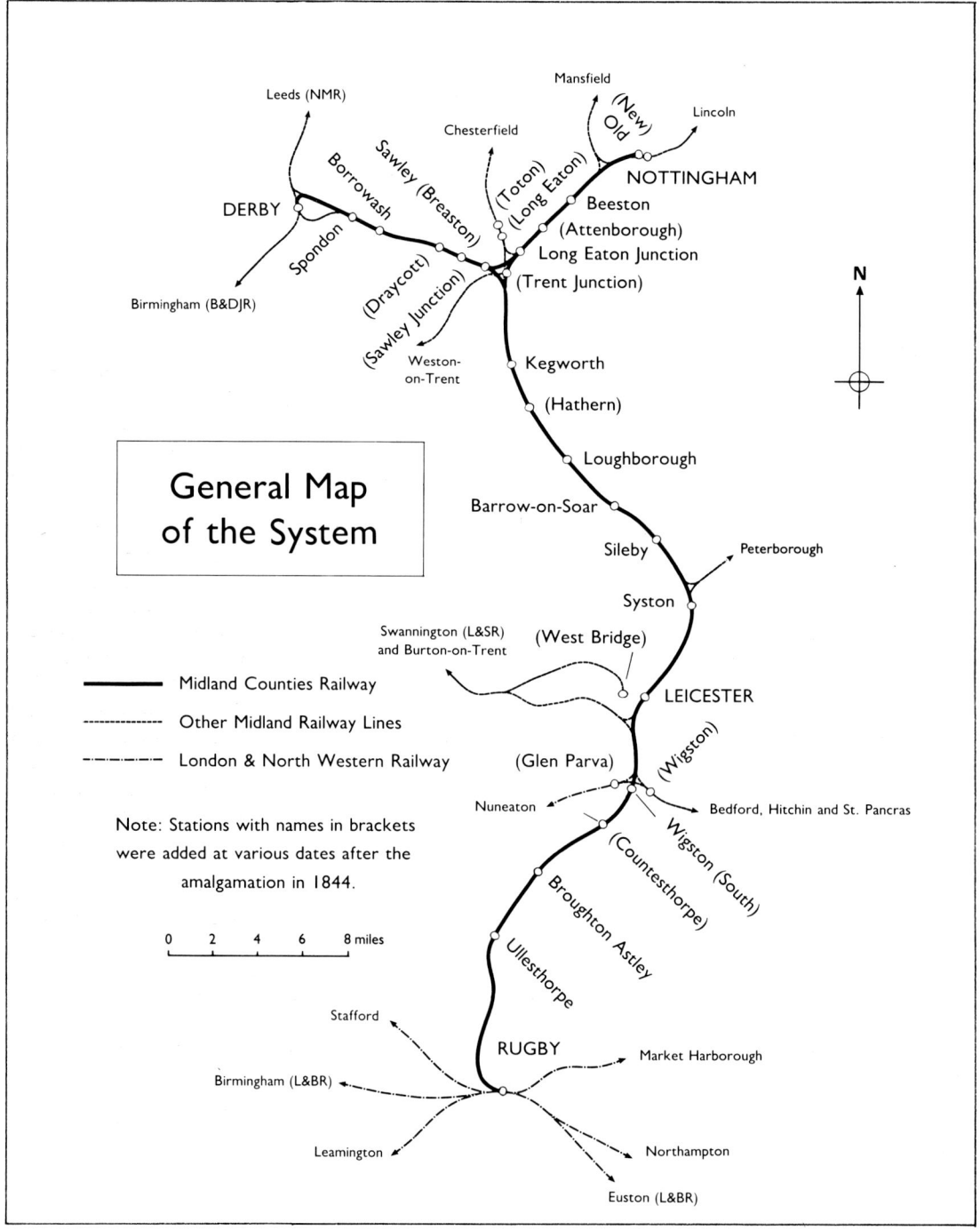

Leeds (NMR)

Mansfield

(New)
Old

Chesterfield

Lincoln

NOTTINGHAM

DERBY

Borrowash

Sawley (Breaston)

(Toton)
(Long Eaton)

Beeston

(Attenborough)

Spondon

Long Eaton Junction

(Draycott)

(Trent Junction)

Birmingham (B&DJR)

(Sawley Junction)

Weston-
on-Trent

Kegworth

(Hathern)

Loughborough

**General Map
of the System**

Barrow-on-Soar

Sileby

Peterborough

Syston

Swannington (L&SR)
and Burton-on-Trent

(West Bridge)

LEICESTER

———— Midland Counties Railway

------- Other Midland Railway Lines

–·–·–· London & North Western Railway

(Glen Parva)

(Wigston)

Nuneaton

Bedford, Hitchin and St. Pancras

Note: Stations with names in brackets
were added at various dates after the
amalgamation in 1844.

(Countesthorpe)

Wigston (South)

Broughton Astley

0   2   4   6   8 miles

Ullesthorpe

Stafford

RUGBY

Market Harborough

Birmingham (L&BR)

Leamington

Northampton

Euston (L&BR)

N

1. General Map of the system

Background and Early Schemes

Although the writing was already on the wall, by 1824 canal investment was still viewed in an attractive light. Towards the end of that year the pages of the *Derby Mercury* were full of reports about two rival projects to link Sheffield with the Peak Forest, Chesterfield, and Cromford Canals. On 1 October, at an adjourned meeting of supporters of the *Grand Commercial Canal,* held under the chairmanship of the High Sherrif at the Great Hotel, Buxton, the prospectus and plan of this scheme were considered. Resolutions passed at earlier meetings in Buxton (on 15 July) and in Sheffield (on 19 August) were read and it was agreed to lend support to the idea of raising by voluntary subscription sufficient money to enable Thomas Telford to report upon this plan, rather than by the immediate issuing of shares. The meeting stood adjourned to the Bridgewater Arms in Manchester on Friday, 29 October.[1] At this next meeting it was agreed that enough money had been pledged to allow the intended approach to be made to Telford. Should a committee be appointed to superintend the rival scheme for the *Sheffield New Canal or Railway* at an intended meeting on 4 November, the supporters of the Grand Commercial Canal recommended putting out an invitation for unification of subscriptions and interests. Thanking Messrs Dean for the plans produced, the appointed committee decided to meet again at the Great Hotel in Buxton on Monday, 15 November, unless Telford's opinion made other arrangements necessary.[2]

Three days before this committee was due to reassemble, another public meeting was called in the Town Hall, Sheffield, to consider the best mode of linking that town with Manchester *by canal or rail road,* and also to communicate with any other canals or railways in the vicinity. It was agreed that the resolution made at the Tontine Inn on 19 August, under the chairmanship of the Master Cutler, should be adhered to: "That the most important advantages would be conferred upon both Town and Trade, and upon the Sheffield Canal, by uniting that canal with the Peak Forest, Chesterfield, and Cromford Canals''; the meeting at Buxton had also taken note of the existence of the railway linking Mansfield with the Cromford Canal at Pinxton. Those present at the public meeting, under the chairmanship of H. Parker, pledged themselves ready cordially to co-operate with the promoters of such canal union in carrying their previous resolutions into effect, and more subscriptions were solicited to pay for the engineer's report.[3] Parliamentary Notice of *Mr Telford's intended Sheffield and Manchester Canal,* from Sheffield to the Peak Forest Canal, was issued by William Tattershall, Solicitor, on 6 November but, although press comment put a very favourable value on the shares in this undertaking,[4] it was pursued no further.

THE CROMFORD AND HIGH PEAK RAILWAY

By this time the attention of subscribers to improved communications had switched elsewhere, the promotion of what was at first termed *the High Peak Steam Railway* having already begun. On Saturday, 5 June 1824, notices appeared in newspapers in Nottingham, Derby, and Manchester announcing a meeting to be held "at the house of Mrs Cummings, The Old Bath, Matlock, on Wednesday, 16 June, at 12 o'clock to consider the expediency of forming a communication between the Cromford and Peak Forest canals by an iron railway, with a branch to Macclesfield, and to take such steps as may be necesssary for carrying the plan into execution". Samuel Oldknow, a noted industrialist in the north-west and a promoter of the Peak Forest Canal, then High Sheriff of Derbyshire, chaired the meeting, which resolved "that no gentleman from Macclesfield having attended the meeting, the consideration of forming a branch railway to that place be deferred" and that no expense should be incurred until £50,000 had been subscribed. At the next meeting, in Manchester on 28 July 1824, it was reported that this had been achieved. The branch to Macclesfield was to be abandoned and Josias Jessop was appointed Engineer and asked to prepare a plan and estimate for the remainder of the project.[5]

Although William Jessop took no direct part in the affairs of the Butterley Company and favoured use of the iron edge railway rather than the plate tramroads popularised by his partner, Benjamin Outram, their continued work as engineers promoted wide use of the company's various products. When Outram died in 1805 Jessop's third son William took over as manager of the ironworks, although he was not actually admitted to the partnership; he was assisted by Outram's brother Joseph until 1815.[6] Jessop's second son, Josias, born at Fairburn in Yorkshire on 24 October 1781, became a civil engineer, having acted as assistant to his father on the building of the Croydon, Merstham & Godstone Railway and on the construction of Bristol Harbour.[7] He had already been responsible for the construction of the Mansfield & Pinxton Railway, was

later called upon to survey, amongst others, a railway between Bristol and Birmingham[8] and, after the Act had finally been obtained, was appointed Consultant Engineer on the Liverpool & Manchester Railway in association with George Stephenson.[9] Before any of these projects were complete, however, he died at the relatively early age of forty-five on 30 September 1826, worn out by exertions in the High Peak.[10]

Josias Jessop's report was presented to the promoters' committee on 2 September 1824, proposing a system of levels and inclined planes, by which stationary steam engines were to be employed with locomotive or travelling engines between the inclines. He estimated an income of £10,000 from 40,000 tons of grain carried 30 miles at 2d per ton per mile, £3,500 from 60,000 tons of coal carried an average 10 miles at 1d and a further sum of £16,676, eleven per cent on the anticipated outlay of £150,000.[11] Parliamentary notice of the proprietors' intention to apply for a bill authorising construction of the railway line was given by Brittlebank & Sons, Solicitors, at the beginning of October,[12] and at a meeting at the Bridgewater Arms in Manchester on 1 December 1824, under the chairmanship of Thomas Bateman, it was reported that the Duke of Devonshire had become a patron of this project, giving consent for the use of his lands. Jessop reported that he considered a railway the best mode of communication between the Cromford and Peak Forest Canals, having estimated the cost of construction at £130,000, exclusive of £20,000 for the stationary engines. Of this sum, £70,000 had already been subscribed in £100 shares, and it was resolved that the remainder should be offered to the public. Brittlebank & Sons' report noted that the meeting stood adjourned to Thursday, 6 January 1825, at the Old Bath Hotel, Matlock.[13] The newspaper also carried Jessop's second report, issued from Butterley Hall on 29 November, commenting that subscriptions had subsequently been found for the full £150,000. It was to prove woefully inadequate in the event, Rennie having already estimated that it would cost £650,000 to form the same connection by canal in October 1810.[14]

Hardly had the line been projected when a meeting was called at the Red Lion Inn at Wirksworth to consider the expediency of forming a branch railway from "Rise End" – presumably at the head of either Sheep Pasture or Middleton inclines – to connect with that town. Assembling again on 8 January 1825, the promoters adjourned to meet once more on the twenty-second of the month.[15] Authorised by Act on 2 May 1825, the Cromford & High Peak Railway was completed by Thomas Woodhouse and opened in two sections: Cromford Wharf to Hurdlow incline foot on 29 May 1830, and thence to Whaley Bridge on 6 July 1831, the traffic being worked by horses between the inclines, of which there were nine, five rising from the Cromford Canal to the summit above Buxton and four falling to the level of the Peak Forest Canal.[16] Of the £164,400 authorised to be raised by the Act of 1825, the shareholders subscribed £127,700 in shares of £100 and £31,910 was raised on mortgage. A second Act had to obtained in 1843 to authorise the raising of a further mortgage of £54,800, of which £22,890 was used to discharge part of the floating debt of £46,915 still owing to the Butterley Company for stationary engines, machinery, rails, and other ironwork on 31 August 1842. This left a second floating debt of £24,025.[17] Could this impecunious and imperfectly engineered concern really have been the target for traffic over such an important project as the London Northern Railroad, promoted at the end of 1824 and still seeking incorporation in 1830?

THE GRAND JUNCTION RAILROAD

The stylised map accompanying the prospectus of this undertaking indicated railways already in contemplation between Bristol and London, between both of those places and Birmingham, thence to Liverpool, from there to Manchester, and also between Leeds and Hull, this last line being isolated from the remainder of the system. The Grand Junction Railroad proposed to connect Birmingham with Derby, Sheffield, and Leeds, bringing the Hull line into communication with the other projects, and intended to throw out branches to Nottingham, from Sheffield to Manchester, and also to the East Coast at Goole. A capital not exceeding £2,000,000, divided into £50 shares, was to be reserved in great measure for the owners of the lands along the route, but with a proportion available for distribution to any of the railroad companies with whose lines the system was to connect. The engineers were listed as James Walker in London and William Brunton in Birmingham.[18] In 1808 Brunton had left the Boulton & Watt works at Soho, Birmingham, where he had latterly risen to the post of superintendent of the engine manufactory, for a post of greater responsibility at Butterley, where he was able to contribute greatly to the success of the new engine plant. When he left to return to Birmingham as managing partner of the Eagle Foundry in 1821, he was succeeded at Butterley by Joseph Glynn.[19] The Board of Management of the Grand Junction Railroad Company was to sit in London, with a group of Country Directors meeting in Derby and professional agents throughout the system, as set out in Table 1.

Application for shares was said to be increasing daily, upwards of 7,000 having already been applied for in London and greatly exceeding the number reserved as the proportion required for landowners. The Board convened in the Committee Rooms at the King's Head Tavern in The Poultry on 24 January 1825, resolving to appoint a committee to examine the proposed line and submit a report.[20] This was received a week later, when it was decided to appoint John Rennie as additional engineer to take the survey of the line between Birmingham and Derby. Josias Jessop was to examine the line from Derby to Wakefield and to take into consideration the practicability of forming a branch to Newcastle-under-Lyne and the Potteries, whilst James Walker was to survey the remainder of the system from Wakefield to Leeds and from Sheffield to Goole and, via

Stockport, to Manchester. William Brunton's part in the organisation would be as manager of the locomotive establishment.[21] To avoid over capitalising, at its meeting on 8 February the Board resolved to ask the engineers to furnish more accurate and definite estimates of the work required, and to postpone distribution of shares until this information had been received.[22] Nothing further is known of the actions of this group of promoters and it is probable that the financial uncertainties

which affected the London Northern scheme had a similar effect upon this one.

THE LONDON NORTHERN RAILROAD

It is with this scheme, noted by Tomlinson as amongst the most ambitious of the period, that we are most directly concerned, bequeathing as it did to the Midland Counties

---

The Board of Management in London
Joshua Walker, Esq., MP – Chairman
Timothy A. Curtis, Esq., – Deputy Chairman
OTHER DIRECTORS

Sir Edward Bankes, Bart.	Moses Montefiore, Esq.
John Bent, Esq., MP	Thomas Meux, Esq.
Mr Alderman Bridges, MP	William P. Lett, Esq.
I. L. Goldsmith, Esq.	John Plummer, Esq., MP
James Farquhar, Esq., MP	Edward Samuel Walker, Esq.

William Howard and William Hill, Esqs., – AUDITORS
Messrs Everett, Walkers, Malby, Ellis & Co, Mansion House – BANKERS
James Wigram, Esq. – COUNSEL
Messrs Allsopp, Parke & Freeth, Lincoln's Inn Fields – SOLICITORS

In the Country, at Derby:
DIRECTORS

Bache Heathcote, Esq.	James Sutton, Esq.
James Holworthy, Esq.	Henry Mosely, Esq.
Robert Newton, Esq.	Francis Sanders, Esq.
William Newton, Esq.	Samuel Rowland, Esq.
John Pares, Esq.	Roger Cox, Esq.
W. J. Lockett, Esq.	John Cox, Esq.

William Strutt, Esq.

Messrs Crompton, Newton & Co – BANKERS
Messrs Balguy, Porter & Barber – SOLICITORS

In Birmingham
Messrs Attwoods, Spooner & Co – BANKERS
Thomas Mole – SOLICITOR

In Manchester
Messrs Heywood, Brothers & Co – BANKERS
Messrs Duckworth & Denison – SOLICITORS

In Nottingham
Messrs J & I Wright & Co – BANKERS
Messrs Allsopp & Freeth – SOLICITORS

In Sheffield
H. Walkers, Eyre & Stanley – BANKERS
Charles Brookfield – SOLICITOR

TABLE I   *Grand Junction Railroad Company Board of Management and Agents*

promoters their first survey. The promoters first assembled under the chairmanship of William Williams at the London Tavern in Bishopsgate Street on 23 December 1824, when it was resolved that a railway connecting London with Manchester, Birmingham, and Hull being of great public utility, a company should be formed with a capital of £2,500,000 in order to bring this into effect. At an adjourned meeting one week later twelve London directors were appointed to serve with the chairman, with full power to co-opt a further seven directors from the districts to be served. Applications for £100 shares, available from the beginning of 1825, were to be made to the Chairman of the undertaking at the Old London Tavern and surveys were to be begun with a view to depositing a Bill in the next session of Parliament.[23] The professional team of agents is set out in Table 2, together with the full list of patrons.

At their meeting on 5 January 1825 the directors resolved that competent persons be immediately employed to examine the two lines between London and Manchester which had been recommended as offering the greatest advantage to both public and subscribers. The first ran from London by way of the Vale of the Lea to Ware and then through Cambridge, Peterborough, and Oakham to a point near Loughborough, whence branches would strike off towards Nottingham and Derby, and from which point the main line would continue until it joined the intended Derbyshire Peak Railroad at Cromford, proceeding thence to reach Manchester via Stockport. The second suggested route ran from London through Northampton, whence there would be a branch through Coventry to Birmingham, and onwards through Leicester and Derby. From Derby there would be a branch to Nottingham and the main line would proceed as in the first plan via Cromford. In addition to its main route the company might build a line from Manchester to Hull, with or without connecting to other undertakings in the vicinity, and a branch from Derby through Sheffield to Leeds. So great was the number of applications for

shares in London that the public was informed that distribution would be deferred until 15 February, offers to be received until the beginning of that month.[24]

Notices issued by the Secretary, R. Mills, on 1 and 12 February, indicated that a surveyor had been instructed to examine the most direct route through Northampton, Leicester, and Derby, the Board having definitely decided upon inclusion of the line from Derby to Sheffield and Leeds whilst continuing to examine which of the other parts of the scheme to adopt.[25] The Board met again on 25 February, under the chairmanship of Pascoe Grenfell, to consider the various engineers' reports upon the alternative routes and opted for that through the valleys of the Lea and Soar. The line would commence by the Thames below London Bridge and pass through the counties of Cambridge and Huntingdon, continuing to reach Manchester by such route "as may on further report be deemed most eligible."[26] On 20 May it was decided that the attention of the company should be confined to linking London with Cromford, there to connect with the High Peak Railroad and from which point connection with Manchester might later be effected. These decisions were communicated to the subscribers at the Old London Tavern on 4 November, when George Hibbert took the chair. The engineers' reports and plans had been completed, apart only from slight deviations which might not be settled in time for deposition of a Bill during the next session. It was still intended that the Bill should be presented in 1826.[27]

The general economy of the country began to decline, however, and many of these early schemes began to founder. A resolution to dissolve the company having been lost at a meeting held on 17 August 1826,[28] it was decided in December to open new subscription books and to issue a second prospectus.[29] Some subscriptions were withdrawn, leaving a capital of only £400,000 upon which deposits had been paid, and it was decided to await better times and the completion of the Liverpool & Manchester Railway. With this now in place and schemes in hand for connections to Sheffield, Leeds, and through Stockport to the High Peak Railroad (already settled upon as the immediate destination of the intended London Northern line) the company was revived towards the end of 1830.[30] Commencing now near the East and West India Docks, the line was to take the relatively level course previously planned through Ware, Cambridge, and Peterborough, to Stamford, where it was to divide into two parts, one continuing through Lincoln, Thorne, and Selby, to York. Uniting in its course with that of the Leeds & Selby Railway, this section would thus effect the most direct communication between Sheffield and York. The second part of the system beyond Stamford was intended to bring the line through Oakham, Melton, Leicester, and Loughborough, on its way towards

---

The Most Noble the Marquis of Lansdown[1] – HONORARY PRESIDENT

HONORARY VICE PRESIDENTS
The Most Noble the Marquis of Anglesea[1]
The Right Honourable the Earl of Hardwick[1]
The Right Honourable the Earl of Lauderdale[1]
The Right Honourable Lord Dacre[1]
The Right Honourable Lord Grenville[1]
Sir Robert Peel, Bart[1]
The Right Honourable Lord Viscount Dudley and Ward[2]
The Right Honourable the Earl of Tankerville[3]
The Right Honourable the Earl Cowper[3]
The Right Honourable Lord Scarsdale[3]
J. George Lambton, Esq., MP[3]

George Hibbert – CHAIRMAN

DEPUTY CHAIRMAN

Pascoe Grenfell, MP	John Smith, MP
Lewis Lloyd	Edward Wakefield

DIRECTORS:

James Evan Bailie	John Irving, MP
Francis Baring	George W. Norman
Lyndon Evelyn, MP	Frederick Pigou
Edward Ellice, MP	Thomas Richardson
Sir Robert Farquhar, Bart	James Warre
Charles David Gordon	William Williams, MP
Sir John Thomas Stanley, Bart[1]	Joseph Strutt, Esq.,[1] of Derby
of Alderley Park, Macclesfield	Ichabod Wright, Esq.[1]
Thomas Markland, Esq.[2] of Manchester	of Nottingham
John Wright, Esq.[3]	Sir Charles Henry Colvile[3]
of Lenton Hall, Nottingham	of Duffield

1 – appointed 12 January 1825
2 – appointed 19 January 1825
3 – appointed 9 February 1825

Richard Hart Davies, MP, Joseph Fry, Simon Mc'Gillivray
and Edward Goldsmith – AUDITORS
Messrs Smith, Payne & Smith, Mansion House Place, and Sir James
Esdaile, Hammet, Grenfell & Scott, Lombard Street – BANKERS
William Vizard – SOLICITOR
Nathaniel Hibbert – STANDING COUNSEL
George Stephenson – ENGINEER

TABLE 2   *London Northern Railroad Board of Management and Agents*

Nottingham, Derby, and Cromford.

Realising that to attempt such a great deal of work at one time would be a hopeless task, the promoters had now decided to proceed piecemeal, going to Parliament in the first instance only for that section of its system between Cromford and Leicester. Returns from the coal and stone traffic expected to make use of this section were regarded as likely to give the proprietors immediate and ample recompense for their outlay, whilst also furnishing construction materials for the remainder of the system. As the pledged subscriptions were felt to be more than adequate for execution of this first section, Parliamentary notice was therefore given for submission of a Bill sanctioning a railway from Cromford to the eastern side of Belgrave Gate in Leicester, near the public canal wharf, and for a branch from Codnor Park to connect with the Mansfield & Pinxton Railway.[31] It is clear from later discussions, arising from their use by the Midland Counties Railway promoters, that the plans and estimates accompanying this application had been drawn up by Josias Jessop at the time of the original promotion of the London Northern Railroad.

Although not needed for the making of this first section, the proprietors intended to raise the full £2,000,000 capital thought to be required for extension of the line from Leicester to London, for which powers would be sought in the following year, and thereafter between Stamford and York. The Directors would be left to apply such of the subscriptions necessary for construction of the first parts of the railway, for a depot near London, and for a sufficient establishment to make the company carriers on the system. According to William Jessop's remarks to the public meeting called to discuss the Grand Midland Railway project, referred to below, the promoters of the London Northern Railroad were now thinking on a fairly expansive scale, with four-track main lines. Their estimates of the likely cost were, however, wildly inaccurate, as events were soon to prove, and the Bill did not proceed.

THE GRAND MIDLAND RAILWAY

Not everyone was convinced that the plans of the London Northern Railroad offered the best means of bringing the East Midlands into communication both with London and Manchester. The Grand Midland Railway originated in a meeting at which a deputation from Northampton suggested that these objects could best be achieved by making a connection from the London & Birmingham Railway, already in contemplation, through Northampton, Leicester, Loughborough, and Derby, to Manchester. At a public meeting in the Guildhall at Derby on 11 November 1830, with the Mayor in the chair, T.P. Bainbrigge read through the proceedings of two previous meetings, at which a provisional committee had been formed to promote these objects. Mr Moss outlined the proposed scheme, explaining that Nottingham would be served by a branch, while the main line was to be taken directly through Derby, as distinct from the route of the London Northern scheme, which was to go via Butterley and come no closer to Derby than the canal bridge at Sandiacre. Nor would the projected railway between Birmingham and Liverpool afford any accommodation to the town, leaving it without any main line of communication.[32] On behalf of that scheme Mr Benton expressed his willingness to confer with any deputation appointed for the purpose and the meeting also listened to the opinions of Mr Gawthorne and Robert Stephenson.

To show that there would be sufficient traffic to warrant the proposed expenditure on a line between Northampton, Derby and Manchester, Moss produced figures showing that the existing coaches and post chaises collected upwards of £293,000 per annum between Northampton and Manchester, with another £192,000 for traffic between Northampton and Derby only. His expectation of costs, amounting to £15,000 per mile, was somewhat greater than that of Messrs Stephenson and Son, at £12,000. With interest and running costs added, an outlay of not more than £1,470,000 should suffice for the line between Northampton and Manchester, or £730,000 for the section between Northampton and Derby only. William Jessop, putting the case for the London Northern scheme, said that when the original surveys had been made in 1824 the costs of construction of a double line had been estimated at £6,000 per mile. This had now been increased by a figure of £8,000 for a four-line route between London and York, and from Stamford through Leicester. At the crossing of the river Trent two passenger lines would diverge towards Derby, leaving two goods lines to pass into the coalfield. He recognised that the High Peak railway had never been intended for passenger traffic and that the best route for such traffic between Derby to Manchester probably lay via Cromford, Bakewell, and Chapel-en-le-Frith, where levels had already been taken.

The comparative costs were therefore £14,000 per mile for the four-track London Northern line against £15,000 for the two-track line between Derby and Northampton. Asked why the Derby branch had been omitted from the deposited plans, Jessop replied that there had been insufficient time to prepare the survey, but that such a branch would be built as soon as the first portion of the line was open for mineral traffic. His personal interest in ensuring that the railway was made as close as possible to his ironworks at Butterley did not go unnoticed, however, and the meeting finally approved of the provisional

committee's exertions and appointed thirteen delegates (listed in Table 3) to confer with all interested parties in order to promote the primary object of the meeting and ensure that Derby would be served by a main line.

A prospectus was issued seeking to raise £1,500,000 in £100 shares upon which a deposit of £3 would be required to accompany all applications. One correspondent dared to question the figures given at the meeting in relation to the revenue for coach traffic. The promoters of the London & Brighton Railway had calculated the revenue from the many coaches plying the 54–58 miles between those places at £73,000, a great variation against the £192,000 supposed to be taken by the three or four proprietors holding the monopoly over the 60 miles of road between Derby and Northampton. The correspondent also wanted to know why no further public meeting had been called to receive the report of the committee in order to instil full confidence in the scheme before the issuing of the prospectus.[33] One of the committee members replied that the prospectus was issued under the sanction of a meeting between the provisional and town committees, at which ten out of the thirteen appointees to the latter were present, their purpose being to allow the inhabitants an immediate chance to support the direct connection of Derby with both London and Manchester; the calling together of another public assembly would be premature and a wasted effort for the moment.[34] Such remarks were described by the first correspondent as reminiscent of a Stock Exchange bubble promotion. He wished to know the names of the provisional committee members, which had not yet been published, and of any eminent subscribers, thinking it extraordinary that the committees should support the issuing of a prospectus before a proper report had been received and published for the benefit and security of potential subscribers.[35] It seems unnecessary to add that this particular scheme was destined to falter, although the inhabitants of Derby were eventually able to ensure that all of the early railways serving their town came together into a single station.

TABLE 3
Grand Midland Railway Derby Delegates

Thomas Wright	D. Fox
William Baker	R. Cox
William Smith	W. J. Lockett
John Wright	Mr Gawthorne
Thomas P. Bainbridge	Dr Baker
Dr F. Fox	Mr Severne
Henry Mozley	

Struggling for recognition

Since the failure of the Leicester Naviga-tion Company's Charnwood Forest Branch during 1799, the collieries work-ing the outcropping seams along the bound-ary between Nottinghamshire and Derbyshire in the Erewash Valley had built up a consider-able and almost monopolistic market for their canal-borne coals in the Leicester area. In 1832 they were selling at a price of 15s per ton, half of which represented the pit-head price and the other half the canal rates, composed of tonnage canal dues varying between 4s 8d and 5s 6d and a carrying charge of 2s 6d.[1] The opening of the first section of the Leicester & Swannington Rail-way on 17 July,[2] however, promised to bring large amounts of the West Leicestershire coals back into the town once again at competitive prices. Although generally considered to be somewhat inferior in quality to those from the Erewash Valley, these were also priced 7s 6d at the pit-head and, being carried at only 2s 6d per ton on the railway, were selling in Leicester at 10s. It was estimated that the market loss to the Erewash collieries would amount to an annual figure of some 160,000 tons, over three-quarters of their entire out-put.[3] In fact, the superior quality of the Erewash coals ensured the retention of this market and it was even extended by 10,000 tons during 1833, although during the same period the Leicester & Swannington Railway had brought in more than 42,000 tons.[4] But to the Erewash coalmasters the situation in mid-1832 looked extremely bleak, and a special meeting of the coal proprietors and lessees was arranged at the George Inn at Alfreton on Monday, 27 August 1832, to dis-cuss the problem.[5]

THE COMPANY IS FORMED

That certain preparations had been made for the Alfreton meeting between 17 July and 27 August is certain. Approaches for a reduction of the canal tolls were reported to have been abortive, although when it was seen that a railway was seriously intended the three canal companies involved in the route between Pinxton and Leicester each offered a 6d reduction, so bringing the tonnages down to between 3s 2d and 4s 0d with the freightage charge remaining at 2s 6d.[6] Williams ascribed the date, 16 August, of one of the group's regular meetings at the Sun Inn at Eastwood as being the actual commencement of these negotiations,[7] but the coal owners' records to which he had access[8] are not at present avail-able and the Alfreton meeting represents the first occasion upon which the scheme was brought to the notice of the general public through the medium of the local press.

Under the chairmanship of Edward Miller Mundy of the Shipley Collieries, the meeting was advised that the best course open lay in the extension of the Mansfield & Pinxton Railway to Leicester, for which very purpose surveys had already been carried out on behalf of the London Northern Railroad between 1824 and 1830. It had then been estimated that £130,000 would be a sufficient sum with which to build a single line for locomotive operation, and that £200,000 would cover a double line. With a cover charge of 6d per ton to defray expenses, it had been calculated that the transfer of the 160,000 tons of coal already passing along the canals would, at 1d per ton per mile, bring in a revenue of 15 per cent, quite regardless of any other traffic which might be expected in general goods and other minerals. Such a reduction of not less than 4s 6d per ton in the cost of conveyance might, it was thought, readily extend the available market as far afield as Melton Mowbray, Oakham, Cambridge, and even Bedford. It was therefore resolved that subscriptions should be invited towards raising the sum thought to be required for a single line and to publicise this undertaking as much as possible.

Opposition to such a scheme was natural, as also the fact that it should originate from the canal companies and from the Leicester & Swannington Railway concern. Joseph Sandars of Liverpool, one of the largest shareholders in the latter and a proprietor with George Stephenson of the Snibston Collieries, immediately suggested that the estimate was far too low, that it would prove totally impracticable to work a single line over 30 miles long, and that the expenses involved in doing so would swallow up the whole of the 3s charges.[9] In the event, the project did cost far more than this initial estimate, at an all-inclusive price of over £30,000 per mile for a double line, but by that time the scheme had changed almost beyond recognition. Sandars also suggested that merely by cutting their prices the Leicestershire Collieries could quite effectively keep Erewash coal out of Leicester and even out of both Derby and Nottingham, a somewhat forlorn hope that was not pursued.

Nothing daunted the committee which had been appointed to superintend the undertaking until its proper incorporation by Parliament met once more at the Sun Inn at Eastwood on 14 September 1832 and it was then pointed out to Sandars and the other pessimists that the promoters hoped to profit from the lessons learned in the underesti-

mation of the costs of construction of both the Liverpool & Manchester and the Leicester & Swannington concerns, and of the "unproductiveness" of the Cromford & High Peak Railway.[10] The £130,000 capital proposed did not cover the purchase of locomotives and wagons, nor the building of the necessary feeder branches, all of which were expected to be provided by the individual collieries. A single line would undoubtedly prove sufficient (as indeed it did on the Leicester & Swannington Railway until 1848, when the centre section of this line was finally doubled under Midland Railway Company ownership on becoming a section of the through route to Burton) unless the traffic should increase drastically.

When the subscribers met once again on 4 October 1832, the previously elected committee was asked to continue to serve the function of a Provisional Committee, with some addition to its numbers, until such time as the Directors and Company Officers had been appointed, its main task being to publicise the undertaking whilst promoting its Parliamentary sanction. On 15 October 1832 it reported from Alfreton upon the steps which had been taken so far, using for the first time the Company's intended title of *The Midland Counties Railway*.[11] This report made it quite clear that, from the start, access had been obtained to the London Northern plans of 1830 and that it was already intended to bring in both Nottingham and Derby by making branches from Long Eaton, a course suggested by a correspondent of the *Derby and Chesterfield Reporter* on 27 September, in order to avoid any great disparity between coal prices in Leicester and the nearer centres of population. The feasibility of extending the line beyond Leicester to join the then projected London & Birmingham Railway at Rugby, or perhaps through Northampton, was also already under consideration. Although the choice eventually fell upon Rugby, agitation on behalf of the Northampton route was to prove troublesome.

Formal notice of intention to apply to Parliament was issued by the first solicitors, Messrs Leeson & Gell of Nottingham, on 9 October 1832[12] and the necessary plans were deposited with the relevant authorities on 29 November.[13] These plans, though unsigned and not credited to any particular person, were almost certainly adapted by William Jessop from those which had been deposited for the section of the London Northern Railway between Cromford and Leicester two years earlier. If these were in turn based upon the survey made by Josias Jessop back in 1824, it is questionable whether enough effort could have been put into bringing them fully up to date in the limited time available since the decision to proceed with the new railway. Instead of the original alignment between Langley Mill and Codnor Park, where the lines to Cromford and to Pinxton were to have diverged, a slightly more easterly course alongside the Cromford Canal had been adopted in order to make an end-on connection with the Mansfield & Pinxton Railway at Pinxton Basin, and the section from Codnor Park to Cromford was omitted. Apart from a slight deviation west of Sutton Bonnington, the routes were otherwise identical, both lines terminating near the canal wharf at Belgrave Gate in Leicester. Inspection of the new plans, together with comparative sections of the Liverpool & Manchester and the Leicester & Swannington Railways, was invited by a notice dated 9 February 1833,[14] and various bankers were appointed to receive the subscriptions in Nottingham, Derby and Leicester. The line was to fall at the rate of twelve feet per mile towards the river Trent south of Long Eaton and then to rise at four feet per mile towards Leicester, the greatest embankment being only ten feet high.

THE MAIN LINE COMPANY

No further progress appears to have been made with the scheme in this original form and when, on 12 August 1833,[15] a meeting was held in Barlow's rooms at Leicester with Matthew Babington in the chair, a different atmosphere prevailed. Babington, a Leicester banker of Rothley Temple, afterwards became first chairman of the company, making visits to London, Liverpool, Manchester, Leeds, and Hull, in order to raise capital. He died on 12 August 1836, less than a month after the company obtained its Act, and did not therefore see the start of any construction work.[16] The character of the line had changed somewhat and the major object now appeared to be the inter-connection of the three towns of Nottingham, Derby, and Leicester, and to link them all to London by way of Rugby and the proposed London & Birmingham Railway. John Fox Bell is reported[17] as having said of this juncture at a later date that "the former Company now wound up its affairs and died" but, as Williams points out, no attempt was made "to disconnect in the public mind" this scheme from the previous one. Babington was joined by Messrs Bell, John Bright, C.B. Robinson, J. Brookhouse (*sic*), and T. Toone on a new committee, elected to confer with similar representatives appointed in Nottingham and Derby. Despite

TABLE 4   *Anticipated Traffic, November 1833*

From travellers by coaches, posting and light goods	77,870
Heavy goods by boats and waggons	8,288
Coal, iron, lead, stone, timber, linen, corn, &c.	16,309
	102,467
Deducting expenses for locomotive power, repairs, &c.	37,250
Net surplus	65,217
By increased travelling and light goods	40,620
Merchandise and heavy goods do.	8,160
Revenue, being 19 per cent dividend on £600,000 capital	£113,997

the fact that it was reported that these three committees met at Loughborough on 1 October 1833,[18] the Derby representatives do not seem to have been appointed until 5 November[19] at a meeting at the King's Head Inn, when Bell gave details of the anticipated traffic, as set out in Table 4.

In his absence, it was explained that Babington had expressed his unwillingness to sponsor the project until he had satisfied himself that these figures were sound, sentiments which he had already expressed at both Nottingham and Leicester. Subscriptions for £60,000 having been raised in Leicester and £50,000 amongst the Erewash colliery owners, a new prospectus was agreed upon by the Nottingham Committee on 5 November 1833,[20] and this was made public during the following month.[21] It detailed the increased capital of £600,000, to be raised in £100 shares, and an anticipated 20 per cent return, the figures quoted above having been slightly amended. The lines north of Leicester were to be completed within two years of the Act, when it was obtained, and the remainder of the line to Rugby would follow in time for the opening of the London & Birmingham Railway. The principal patron of the scheme was named as the Rt. Hon. Viscount Melbourne, with a supporting Provisional Committee of 37 members (set out in Table 5) made up to represent the three counties involved. George Rennie and William Jessop were named as Engineers and John Fox Bell Secretary.

Notice of intention to apply to Parliament was issued once again on 12 November,[22] and fresh plans deposited at the end of the month.[23] Between Pinxton and Leicester the route was exactly as before, but the line now extended through the eastern side of the town and along the river Soar as far as the Blaby-Desford Turnpike at Whetstone, pending completion of the survey to Rugby. Between Nottingham and Derby the line was to take the route finally adopted at each end but, between Beeston and Draycott, it lay farther north, intersecting the main line on that side of Long Eaton. There were to be fairly sharp

curves in all four directions at this point, particularly severe in the case of the two from the north. Alternative termini were shown in Nottingham at Castle Gate and London Road and in Derby at Darby's Yard and Exeter Gardens. Rennie reported on 27 November that he had examined Jessop's survey and had been over the complete route.[24] He thought the line selected very favourable, that it could be constructed at a comparatively moderate outlay, and that only a small amount of mechanical power would be required for each of the different sections.

The voice of dissent was heard once again on 20 January 1834 when, under the pseudonym *Observer*, James Clifford of Shardlow, extensive wharfinger and a proprietor of the Soar Navigation, sought to discredit Jessop's survey by referring to the poor estimates for the Cromford & High Peak Railway, which had been exceeded by 35 per cent upon con-

TABLE 5   *Provisional Committee, December 1833*

FOR LEICESTERSHIRE, MESSRS	
C. W. Packe	T. Toone
M. Babington	W. Hackett
T. E. Dicey	C. B. Robinson
W. Heyrick	A. Burgess
J. Bright	Dr Hill
J. Brookes	Col Cheney, CB
J. Needham	

FOR NOTTINGHAMSHIRE, MESSRS	
L. Rolleston	W. Trentham
J. Musters	T. Barber
J. Wright	S. Parsons
J. Coke	R. Renshaw
F. Wright	H. B. Campbell

FOR DERBYSHIRE, MESSRS	
E. M. Mundy	S. Fox, Junr.
W. P. Morewood	H. Chapman
W. R. Newton	J. Wright
J. Boden	J. Sandars
E. Cox	W. Baker
J. Oakes	Byng
D. Fox	Tunnicliffe

struction.[25] In fact of course, those estimates had been prepared in 1824 by his late brother Josias and not by William himself,[26] a fact of which Clifford was either ignorant or which he chose deliberately to ignore. He went on to suggest moreover that the whole Midland Counties project was Jessop's idea alone and that, with the support of a few other coal-owners, Rennie's name had merely been appended to add an air of importance. Rather than have this attack upon his professional character reflect adversely upon the project, Jessop resigned his post on 28 February.[27] The resignation was reluctantly accepted by the General Meeting of Subscribers at the Three Crowns Hotel in Leicester on 3 March,[28] when the attack was deemed most unjust and completely unfounded, and Jessop accepted the offer of a directorship in place of his position as engineer.

Charles William Packe was in the chair at this meeting, which had been called by the Provisional Committee to receive a progress report. So far, deposits had been received on only £124,500 part of the intended capital, whereas it had been expected that a much greater amount would have been forthcoming, and there must consequently be another year's delay before any further progress could be achieved. Coal consumption, however, was on the increase and, although more than 42,000 tons had been supplied in the past year by way of the Leicester & Swannington Railway, the amount brought from the Erewash coalfield to Leicester had exceeded by 10,000 tons the quantity carried in 1832. Postponement of the application to Parliament would enable the Company to complete the survey to Rugby and to obtain an Act for the whole line at one go, thus dissipating idle rumours that its only object was still to bring coal to Leicester. This was now "comparatively a subordinate consideration", the conveyance of passengers and goods between London and the North and Midland districts being much more important and a more reliable source of income. It was also suggested that an extension to Sheffield might be possible, with a ruling gradient of 1 in 330, the greater part not exceeding 1 in 1,200. A directing committee of 22 members (set out in Table 6), of which any five would constitute a quorum, was appointed to complete the survey and to publicise its progress, to consider the desirability of the Sheffield extension, and to appoint another Engineer to act with Rennie.

A further meeting was held at Parker's office in Sheffield on 14 March 1834[29] to appoint a committee especially to consider the Sheffield extension, which would later prove such an embarrassment. The Leicester–Rugby survey was completed by Frederick Simpson in April and it confirmed Rennie's favourable report.[30] A tunnel, one mile long, would however be required at Ashby Parva. The Directing Committee was to meet again at the Bell Hotel in Leicester on 6 August.[31]

With confidence in the project firmly re-established, the canal companies became increasingly disturbed. Too late now, on 27 August 1834 the Soar Navigation decided to reduce its tonnage to 7d, bringing the Leicester price of best Shipley coal down to 10s.[32] It was immediately pointed out that Derby still paid 11s and 12s for the same article, even though much nearer to the source.[33] A special meeting of the Erewash Canal Company was called at the White Lion Inn, Nottingham, on 30 January 1835 to consider what action it should take,[34] and the Derby Canal Company soon followed suit.[35] The London & Birmingham Railway's Liverpool

TABLE 6   *Directing Committee, March 1834*

C. W. Packe	T. Barber
W. P. Morewood	J. Bright
M. Babington	W. Stevens
J. Coke	W. Trentham
F. Wright	S. Parsons
R. Cheslyn	H. B. Campbell
S. Thompson	R. Renshaw
J. Horsefall	W. Hackett
J. Brookes	T. Toone
W. Jessop	J. Palmer
J. Oakes	C. B. Robinson

subscribers meanwhile urged their own directors to lend support to the Midland Counties project,[36] and beds of coal at Pinxton and South Normanton were offered for sale on the strength of its completion.[37]

Notice of the promoters' intention to go to Parliament was issued once more on 4 November 1834,[38] and the revised plans were deposited at the end of the month.[39] This time, in addition to repeating the previous alignments, they contained the remainder of the line beyond Whetstone, taking a direct course through Cosby, Broughton Astley, Ashby Parva, Lutterworth, and Churchover, and entering Rugby along the east side of the river Swift to make a triangular junction with the London & Birmingham Railway on the site of the present station. The curves at the intersection of the Pinxton-Rugby and Nottingham-Derby lines were this time even more ridiculously tight than before.

Once more, however, the Bill was not introduced to Parliament and instead, in February 1835,[40] another prospectus was issued, this time claiming as patrons the Earl of Denbigh, Viscount Melbourne and Lord Scarsdale. Rennie was still, at this date, named as sole Engineer. On 6 August 1835 the Mayor of Nottingham convened another meeting in the Town Hall, chaired by Thomas Wakefield, at which Babington explained the position.[41] Pledging its support, the meeting appointed thirteen representatives (set out in Table 7) to co-operate with the directing committee. The project was at last reported completely subscribed on 2 September 1835,[42] in excess of the amount required by parliamentary standing orders; £60,000 had come from Liverpool and not less than £320,000 from Manchester.[43]

A FRESH SURVEY

From the time in 1825 that he was chosen by John and George Rennie to make the successful survey for the Liverpool & Manchester Railway, Charles Blacker Vignoles was continually involved in the field of civil engineering, both at home and abroad.

Before that appointment he had done military service and a significant amount of surveying in central America. Throughout his life he never lost his sense of impetuousity and, while not alone in falling foul of George Stephenson, was nevertheless able to convince the Lancashire railway subscribers of his worth. Theodore Rathbone, vice chairman of the North Union Railway, had now joined the Board of the Midland Counties Company, together with James Cropper, and at their invitation,[44] during the first week of September 1835, Vignoles was instructed by the chairman to re-examine the directions and levels of the proposed line. In fact he made a complete re-survey, his report being dated 12 November[45] and the new plans 28 November 1835.[46]

Vignoles suggested a total alteration in the course of the line between Rugby and Leicester and an equal change in the section, to obviate the long tunnel at Ashby Parva and the sharp rise of 21 feet per mile from Blaby to Leicester. Collateral advantage was to be gained from these diversions in keeping entirely away from the property of an influential landed proprietor at Brownsover, immediately north of Rugby. The line was now to run west of the river Swift through Ullesthorpe, Broughton Astley (where it intersected the original course), Countesthorpe, and Wigston. Gradients were now to be 1 in 440, rising for six miles southwards from Leicester, 1 in 400 for a further five miles, and thence to the summit at seven feet per mile, afterwards falling to Rugby at 1 in 400. The tunnel was thus rendered unnecessary with only a little additional earthwork and

TABLE 7 *Nottingham representatives, August 1835*

Bean	Needham
Felkin	Parsons
Hall	Pearson
Hanney	Rogers
Hawksley	Trentham
Horsfall	Wakefield
Milnes	

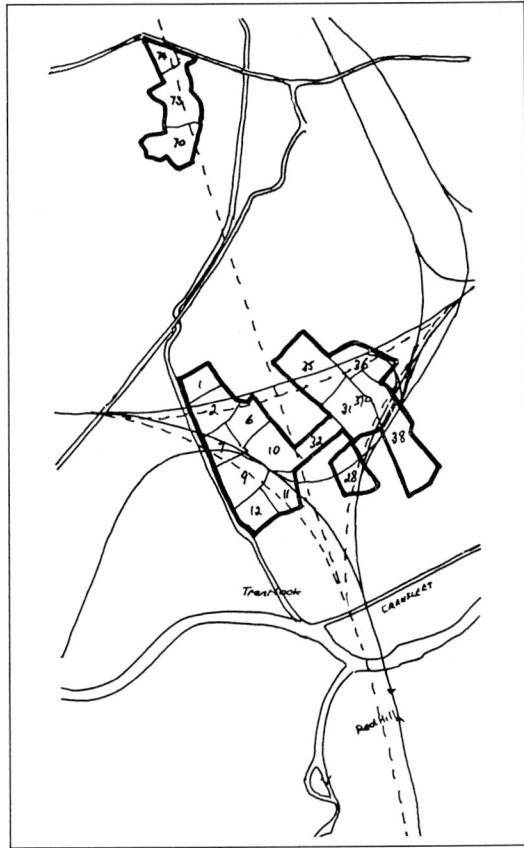

2. Plan of the Trent Triangle
John Chamberlain Hopkins, who carried on a long running campaign to have the railway diverted away from his house at Sutton Bonnington, also owned a large amount of land standing in the way of Vignoles' "capricious" expanded curves south of Long Eaton. Although concerned only with those lands required for railway construction – and Hopkins might conceivably have owned property standing in the way of any route through Sawley – the Parliamentary plans submitted by Vignoles in 1835 might well have been construed as a personal vendetta. The line into the Erewash Valley, which was to be struck from the Bill, also picked out some isolated holdings belonging to Hopkins. They are shown here with the routes of the various railways made down the years, the lines of the Midland Counties company having been built a little farther to the north and east than originally planned. Based upon information from the Nottinghamshire Records Office.

stone required instead. The deep cutting through his estate would not intrude upon the privacy of a certain noble Lord and so remove any parliamentary opposition from that quarter.

Various alterations in both course and gradient were suggested between Leicester and the Trent in order to maintain a uniform gradient. The line was to pass farther east than previously, away from the Soar through Syston and Sileby and with alternative routes through Barrow-on-Soar, but it lay along the earlier route through Loughborough with only minor variations and the introduction of alternative routes through Sutton Bonnington and Kingston-on-Soar. Gradients on this section were to be five feet per mile from the Trent to a point two to three miles south of Loughborough and six feet per mile thence to Leicester, the better engine working easily repaying the additional outlay required for earthworks, in particular the long embankment across the Loughborough meadows. Various changes were also advised between the Trent and Pinxton, the line to pass farther west through Sandiacre and Ilkeston than previously intended and so avoiding intrusion upon Lady Warren's grounds and repeated crossings in the former alignment over the Erewash Canal. The short, deep cutting at Sandiacre would repay its expense by providing good quality building stone. A small deviation would be made between Codnor Park and Pinxton, where the line was to be level, falling thence to a point just south of Langley Bridge at 1 in 400 and uniformly thence to the Trent at twelve feet per mile. In the event, the Erewash Valley section was to be resurveyed on three further occasions before being built on a somewhat different alignment in 1845–47.

Considerable alteration to the Derby-Nottingham line was advised, to widen the curves connecting both east and west with the main line in view of the expected "Great North Route" passing towards Derby, whilst the future connection with Lincolnshire would be made through Nottingham. The wide sweeps in a triangular space between Long

Eaton, Sawley, and Attenborough might appear capricious, but the complicated connections of levels and regard to the passage of the railway over canals and the turnpikes had led to their adoption. The line was therefore brought southwards from its earlier course between Beeston and Draycott, so that the intersection was now south of Long Eaton. From Nottingham the rise would be scarcely more than two feet per mile as far as the junction with the main line, rising thence to Derby at six to seven feet per mile to suit the road and canal crossings and the connecting curves. The termination in Derby being not yet clearly decided upon, two alternative routes were suggested, one north and the other south of the canal, but both ending near St. Mary's Bridge, with a short branch

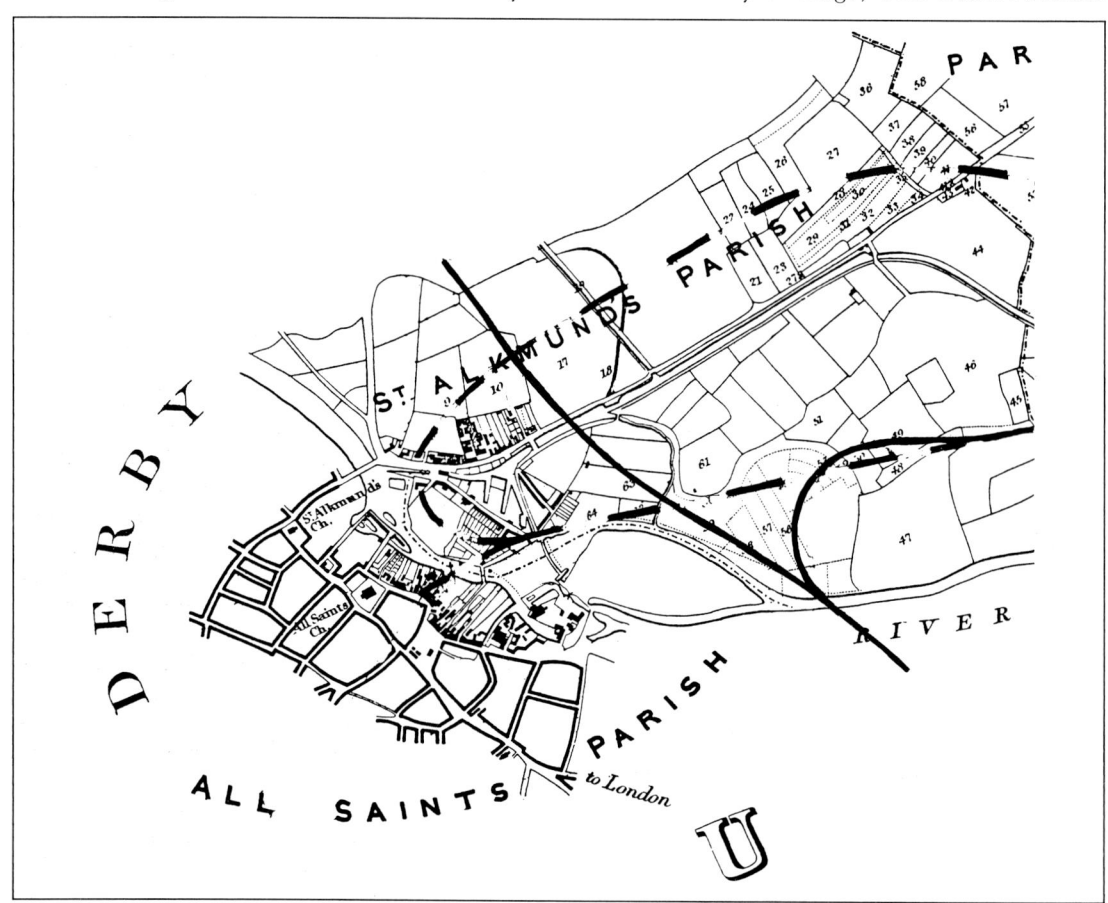

3. Original site plans for Derby Station

In their individual submissions to Parliament the three companies planning to enter Derby had each proposed to build a station in the vicinity of the Holmes, an island formed between the natural course of the river and the head of the original Derwent Navigation. The Midland Counties company had first, in 1833 and 1834, aimed alternatively at Darby's Yard, near the Market Place, and Exeter Gardens, on the north side of the river. Vignoles effectively retained the approach towards the Market Place via Full Street with alternative routes from Chaddesden south and, through the New Pastures, north of the canal to St. Mary's Bridge. The routes of the lines finally built are shown by way of comparison. Based on information from the Derbyshire and Nottinghamshire Records Offices and from the files of *The Derby Mercury*.

to Full Street. Alternative termini were provided for in Nottingham, in Wheeler Gate and London Road. Other small branches which it was thought might prove necessary were a second connection with the London & Birmingham Railway at Rugby facing towards Birmingham (although this curve was not drawn out in the parliamentary plans as first submitted), a branch to join the Leicester & Swannington Railway from the south near the Fosse Road Bridge, two lines in Leicester to a depot near St. George's Church, a branch to the granite quarries at Mountsorrel, and one into Loughborough from the south.

THE BILL IS FINALLY PROMOTED

Contrary to accepted statutory requirements, the Parliamentary Notice of 12 November 1835 appeared in only one of the local newspapers, *The Nottingham and Newark Mercury.* It was not featured in *The Derby Mercury, The Derby and Chesterfield Reporter, The Nottingham Journal,* or *The Nottingham Review,* but this does not seem to have prejudiced the Bill in any way. This, as noted below, was not the first time that the promoters had chosen to advertise their intentions in the less popular press. The notice which did appear and the plans deposited on 30 November both describe the line as set out in Vignoles report (apart, that is, from the Birmingham curve at Rugby, which was omitted at this stage). On 4 February 1836, at the Swan Inn in Mansfield,[47] the Mansfield & Pinxton Railway's Committee agreed that the Midland Counties line should join its own at the Pinxton Wharf terminus and that, in consequence, new rails would be required throughout their line, whilst the road itself needed widening. As events turned out, it was left to the Midland Railway Company to undertake this reconstruction in 1847–49.

When it became generally known that the company actually proposed to make the extension to Chesterfield on its own account, and so cut out a large section of the North Midland Railway from the through route to the North, serious opposition was aroused. A public meeting in Derby Town Hall on 2 December 1835 resolved to oppose very strongly any such measure as being against the town's best interests.[48] Edward Strutt, MP, however, supported the extension, while Sir George Crewe of Calke Abbey expressed a preference for the North Midland route.[49] In a letter to the press on 18 April therefore,[50] the secretary explained that a paper detailing this scheme, which had been published on 11 February 1836,[51] had not been intended as an act of hostility to the North Midland Railway and that, having been thus misconstrued, it had been suppressed by the Board on 11 March. The inhabitants felt, nevertheless, that the company had obtained their assenting petition on 12 March under false pretences and Strutt therefore presented a further dissenting petition in the Commons on 13 April. It contained 2,000 signatures obtained within 24 hours. A further meeting in the Town Hall on 12 March had expressed strong opinions in opposition to any such extension.[52]

It was eventually recommended that the levels of the Pinxton branch should not be raised above the sections already displayed, so presumably rendering any extension towards Clay Cross and Chesterfield that much more difficult, and that this branch should be built as a single track only. It was not expected to carry many passengers and should not cost more than £146,802 17s, including ten per cent for contingencies, the major traffic anticipated upon it being 188,000 tons of coal and 25,000 tons of other minerals, assuming that the railway would share the available traffic equally with the canals. On the remainder of the system traffic was expected to top 466,000 passengers and 198,000 tons of goods, producing revenue amounting to £129,938 and £49,482 respectively. Of the total £785,700 capital subscribed at 1 March 1836, only £299,700 had come from 270 local shareholders. The remaining £486,000 had been pledged by 231 outsiders.

The first moves towards the single station in Derby now proposed for the three railway companies seeking to reach the town had been made at the Council Meeting on 2

February,[53] when a Committee was set up to investigate the matter. Johnson reported on 9 April that with Sandars he had been favourably received by various Directors.[54] When George Stephenson, for the North Midland Railway, had approved of the site suggested on the Holmes, the Midland Counties Bill was altered accordingly, but it had been changed again when no agreement had been achieved over the necessary land.

Negotiations had meanwhile been in progress to divert the course of the line south of Leicester in order to make the point of junction with the London & Birmingham Railway at Northampton.[55] It may be remembered that, back in 1830, the purpose of the Grand Midland Railway was to bring Manchester, Derby, and Northampton into direct communication with one another and with the London & Birmingham Railway near Blisworth. Thomas Howes, a Northampton solicitor, had spoken at the public meeting in the Guildhall at Derby and Robert Stephenson had submitted a report to the promoters. Early in 1832, when he was giving evidence before the parliamentary committee examining the London & Birmingham company's first Bill, Stephenson agreed that it would be feasible to construct a branch through Northampton to Derby and thence

4a & 4b. Original site plans for Nottingham Station

Vignoles plans provided for alternative termini in Nottingham at London Road and, near the Market Place, in Wheeler Gate. Modified during its course through Parliament, the Bill finally provided for the station to be made alongside an extension of Carrington Street. Nottinghamshire Records Office.

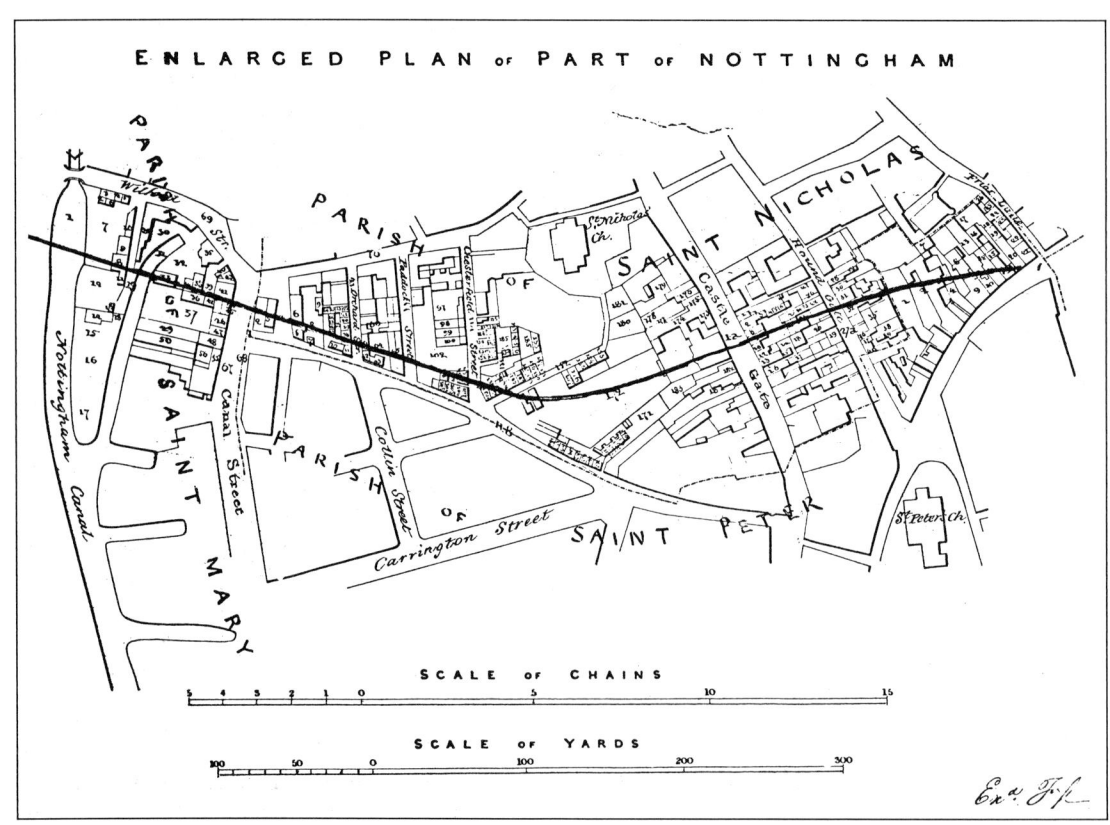

to Nottingham.[56] In their first prospectus, towards the end of 1832, the Midland Counties Railway promoters indicated their intention to make "an extension of little more than 30 miles of railway, southward from Leicester" to connect with the main line of the London & Birmingham Railway. William Jessop had suggested that this could be accomplished "by an extension of only 18 miles to Rugby, or by another and *perhaps more preferable* route of 33 miles to near Northampton". A map which was included with the prospectus showed the two alignments.

With the influx of the Leicester interests during 1833 a choice was made in favour of the junction at Rugby, being more suitably placed to provide an adequate link to Birmingham than any connection farther south. The line to Northampton and Blisworth would have been an expensive one to build across the Northamptonshire Uplands and the valley of the river Nene; considerably longer than the line to Rugby, its use would have saved only four miles on journeys between Leicester and London. The disclos-

ure of this alteration in the plans of the Midland Counties Railway was not immediately followed by protests from the inhabitants of Northampton, possibly because the company chose to insert its advertisement in the least read of the three local newspapers. It was published in the two succeeding issues of *The Northampton Free Press* on 7 and 14 December 1833. Already in decline and a newspaper that "not one person in a thousand ever saw", it went out of business during the following year. Whether the promoters were being deliberately evasive in choosing to patronise such a poorly supported organ is unclear, although they were hardly unaware of the competition. Thomas Edward Dicey, of Claybrooke Hall, Leicestershire, proprietor of *The Northampton Mercury* and a former banking partner of Matthew Babington, owned 2,000 of the railway company's shares, had been an original member of the provisional committee for Leicestershire, was appointed one of the first 24 directors, and eventually succeeded Babington as chairman.

Whatever the merits of the case, this expedient of placing an advertisement in an obscure newspaper succeeded beyond all reasonable expectation. Nobody of consequence in Northampton took any notice or learnt about the changed plans of the Midland Counties company from any other source. Not until 16 January 1836, only a month before the Bill for the Midland Counties Railway was given its first reading in the House of Commons, did the other two newspapers carry the information that the railway from Leicester would join the London & Birmingham at Rugby. The inhabitants of Northampton were at last aroused to activity by this unexpected news. On 4 February a public meeting was held in the Guildhall, at which it was agreed unanimously that local interests would be greatly benefitted if the Midland Counties Railway were to run by way of Market Harborough and Northampton. A committee was formed to consider the best mode of attaining this object, to communicate with the directors of the railway company, and to appeal for sup-

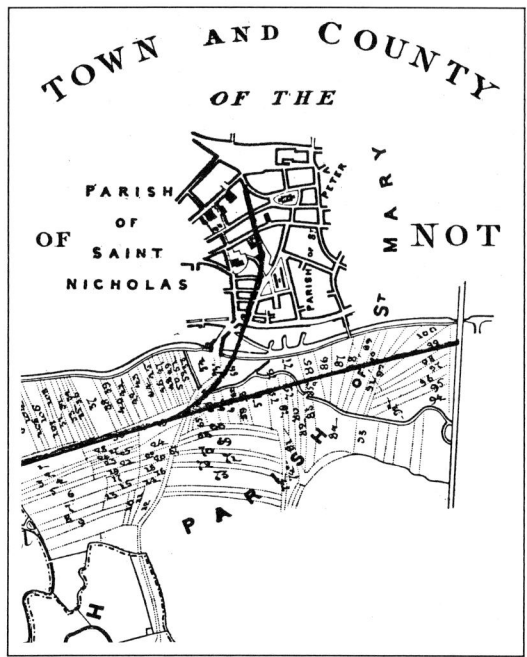

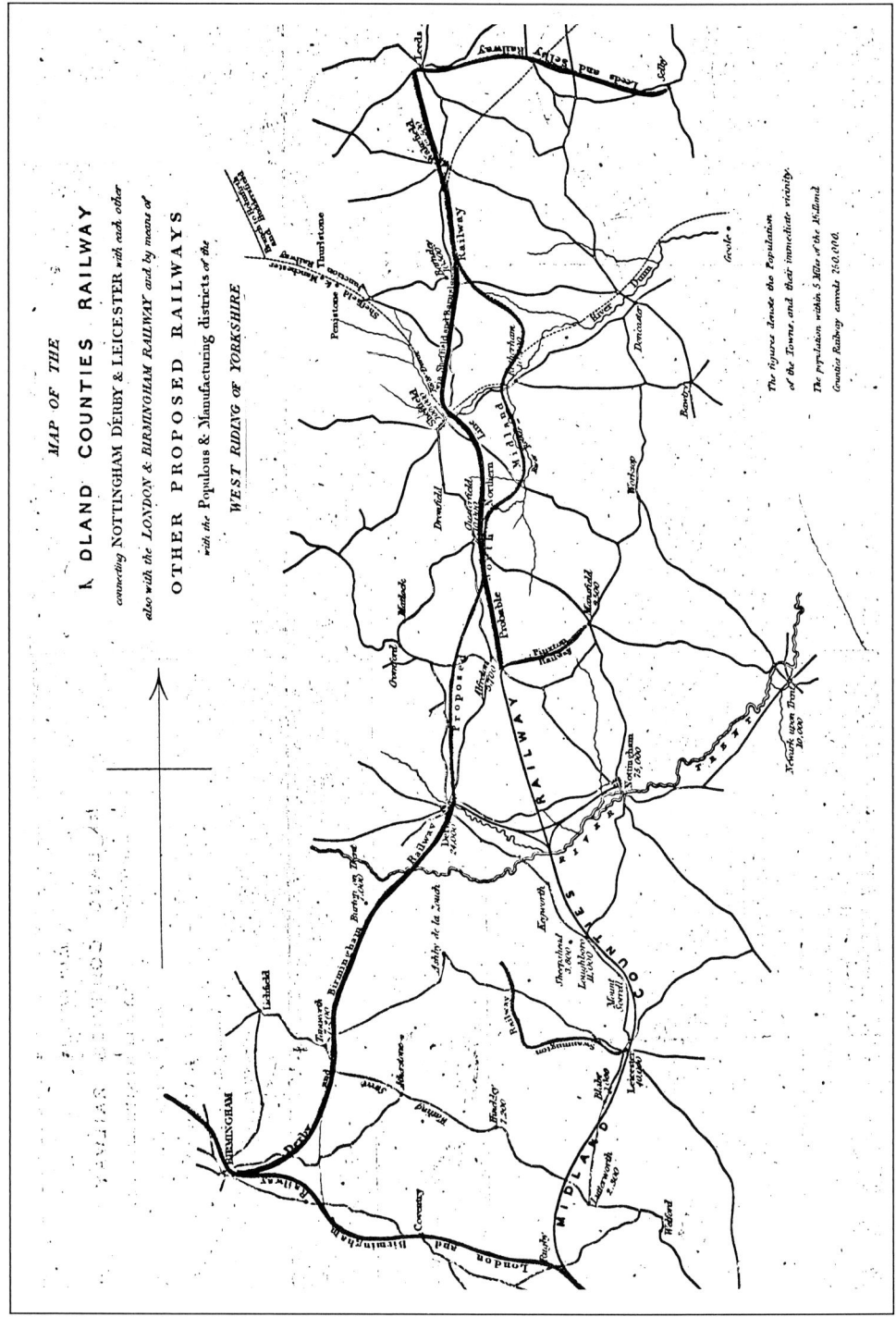

5. Map of the Sheffield and Barnsley Extension

The plan circulated in February 1836 which detailed the *Probable Northern Line via Sheffield and Barnsley*, to which the North Midland promoters took such great exception. It crossed their line – which ran to the east through Rotherham – south of Chesterfield. Nottinghamshire Records Office.

port from the district through which the line was intended to be made. Vignoles reported that the alternative line would cost £500,000 more to build than that to Rugby. Leicester Borough Council, meeting on 8 February, heard one of the Percivals, bankers in Northampton, and an Engineer, Bevan, explain their proposal, and resolved to oppose the Midland Counties Bill as it stood.[57]

In March 1835 it was reported that there were 1,962 petitioners in favour of the Bill, 408 against it, and 330 neutral towards it.[58] During the second reading in the House of Commons it was again suggested that the line should pass through Market Harborough and Northampton, a route six miles shorter than that via Rugby,[59] and on 22 March petitions dissenting from the Bill in its existing form were received from Market Harborough and other places in North Leicestershire, as well as one from Richard Arkwright.[60] The Commons Committee's report was considered on 3 May,[61] and the Bill was recommitted by 178 votes to 25 on the motion of Sir R. Ferguson, supported by Strutt, William Yates Peel, and Sir Oswald Mosley, on behalf of the town of Derby, the North Midland Railway Company, and the Birmingham & Derby Junction Railway Company respectively, on the grounds that the directors had not acted in good faith towards these bodies and were practising evasion by promoting the Chesterfield extension as an independent project.[62] Ross and Vernon Smith, MPs for Northampton, and Finch, MP for Stamford, supported the motion because the Committee had refused to entertain the Northampton case when its promoters failed to deliver plans.

On 5 May Gisborne objected to Robert Palmer's dissenting petition being presented,[63] and moved that the Committee consider only the merits of the Northampton line, the others objecting unsuccessfully against this.[64] The Commons Committee therefore sat on 9 May to examine the engineer, Francis Giles, on the proposed lines through Market Harborough to Northampton and Stamford. It reported that future discussion should be limited to the relative merits of the Northampton and Rugby lines, estimated respectively at £1,000,000 and £300,000, the choice resting chiefly upon whether the Stamford branch was to be made. In order to placate the opposition and ensure an easier passage through the House of Lords, the promoters agreed to insert a clause in the Bill forbidding any action upon the Leicester–Rugby section until 1 August 1837, or the end of the Parliamentary Session then sitting, thus giving the South Midland Company time to apply for its line and then to effect a fusion of interests if it so desired. No further opposition would be made against the current Midland Counties Bill or against any future Bill for the South Midland which might be introduced during the next two sessions of Parliament.

The Bill was read a second time in the Commons on 31 May[65] and a third time on 1 June,[66] Royal Assent being given to the Midland Counties Railway Act, *6 Wm.IV cap.78,* on 21 June 1836. By its means Matthew Babington, John Bright, Robert Benson, Birley, Edward Cropper, Thomas Edward Dicey, George Carr Glyn, Lawrence Heyworth, William Jessop, Henry Jephson, Joseph Frederick Ledsham, Daniel Ledsham, Viscount Melbourne, the Duke of Newcastle, James Oakes, Charles William Packe, Theodore Woolman Rathbone, Thomas Toone, George Walker, and others were incorporated proprietors in *The Midland Counties Railway Company.*

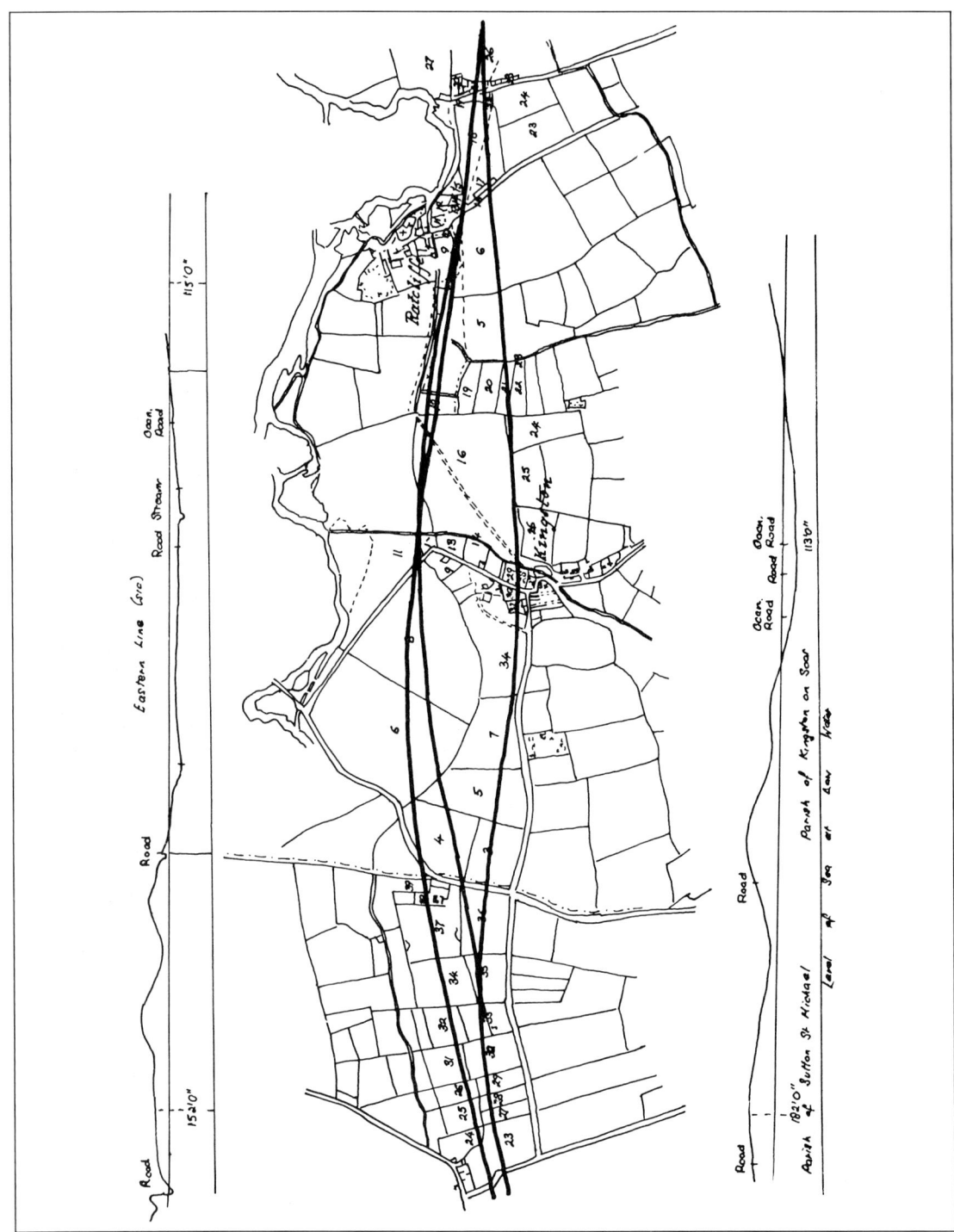

3

Construction and Opening

The Midland Counties Railway Act of Incorporation[1] brought into being a public company whose £1,000,000 capital had to be fully subscribed before any of its statutory powers could be invoked, only £700,000 of the stock having been allocated at that time. An additional £333,333 (the usual one third excess) might be raised by mortgaging the concern. As the Pinxton section had been abandoned, in order to save the remainder of the Bill, construction powers were granted only for the lines between the London & Birmingham Railway at Rugby through Leicester and Loughborough to the Cranfleet Cut (an artificial channel belonging to the Trent Navigation Company in the Parish of Sawley, south of Long Eaton); between the parishes of St Mary, Nottingham, and St Alkmund, Derby; the two connecting curves from the Cranfleet Cut to Long Eaton and to Sawley; an additional north-west curve into the London & Birmingham Railway at Rugby; and two branches to a terminal station in Leicester between Rutland Street and Yeoman Street. Since a compromise route between the alternative lines which had been set out at Kingston-on-Soar was also specified, amended plans showing the authorised routes and duly certified by the Speaker of the House of Commons were lodged by order of Parliament on 18 August 1836.[2]

No more than 22 yards width of land might be taken, except for embankments, cuttings and other building works, the permitted deviation from the authorised centre-line being limited to 10 yards only in built-up areas and 100 yards elsewhere, except in the Parish of Sutton St Michael unless with the landowner's consent. Apart from this last item, inserted into the Act to protect the grounds of a residence in Sutton Bonnington, these were fairly normal terms. On the Kingston – Kegworth road any station must be built so as to occupy not more than one half-acre. Seven years were to be allowed for completing the works, but only two for land purchase negotiations, the money paid for common land in the Nottingham Meadows to be disposed of as decided by a public meeting of at least 2,000 of the burgesses entitled to pasture rights thereon. As with the canals in earlier years, landowners and occupiers concerned in mining might lay down private branches to connect with the railway, though the line hardly penetrated the active coalfield in its authorised form.

Leicester Borough Council secured the several restrictions advised by its Railway Committee on 13 January 1836. Bridges must be

6. Alternative routes through Kingston-on-Soar. The compromise route by which Vignoles hoped to satisfy the opposing claims of landowners at Kingston-on-Soar and Sutton Bonnington, superimposed upon the alternative alignments which had been set out in the Parliamentary plans. John Chamberlain Hopkins owned the house in plot no.38 and Edward Strutt the hall at Kingston. As built, the railway ran, north to south, from top right to bottom left. Nottinghamshire Records Office.

built to carry the New Walk, Princes Street, and an extension of Regent Street across the line and to take the railway over an occupation road which would eventually form Lancaster Road, between Regent Street and the Welford Turnpike. The Company must also build a wall six feet high and not less than fourteen inches thick between this last road and the Regent Street bridge, and make a tunnel under the Freemans Piece.

Until the first General Meeting of Proprietors, to be held within three months of the passing of the Act (and thereafter annually in May or June), the 24 Directors set out in Table 8 were to act for the Company. Appointments confirmed by the Board on 12 August 1836 were Thomas Edward Dicey as Chairman, James Oakes as Deputy Chairman, John Fox Bell as Secretary, John Mansfield as Treasurer and Charles Vignoles as

TABLE 8 *Statutory Directors acting until the first General Assembly*

Matthew Babington
Charles William Packe
Joseph Frederick Ledsham
William Wilberforce Pearson
Francis Wright
Joseph Smith
William Jessop
Theodore Woolman Rathbone
Edward Cropper
Charles Stewart Parker
Richard Cheslyn
Edward Miller Mundy
Lawrence Heyworth
Edward Dawson
Thomas Edward Dicey
John Coke
John Horsfall
James Oakes
Joseph Walker
George Barker
John Bright
Thomas Toone
William Hackett
Joseph Cripps

Consulting Engineer, with T.J. Woodhouse as his assistant.[3] The initial Directors named in the Act had been replaced by another twenty-four, qualified as owners of ten shares in their own names and elected earlier in the day at the first statutory meeting of the proprietors in the Bull's Head & Anchor Inn at Loughborough, a maximum of £1,000 being allowed for their remuneration.

EXTENSIONS TO NORTHAMPTON AND SHEFFIELD

In deference to the promoters of the Northampton line, the Midland Counties Railway Act stipulated that no action must be taken with respect to the portion of line from Rugby to the boundary between the parishes of Wigston Magna and Knighton until 1 August 1837 or, if Parliament should still be sitting, not until the last day of that Session. A Bill for the South Midland Counties Railway received a first reading in the Commons before it was handed over to the attentions of a parliamentary committee investigating railway subscription lists. The report of this committee killed the Bill by revealing that many names on the subscription list belonged to persons of no financial standing. There can be no reasonable doubt that the purpose of this promotion was not to raise the large capital required for constructing a line, but to coerce the Midland Counties company to make its line to a junction with the London & Birmingham Railway near Northampton.[4]

Referring to the Pinxton line in his speech to the proprietors at the General Assembly, the Chairman stated that the Directors had no intention of making any further applications to Parliament, but they would give full support to the independent company which was proposing to build a line to Sheffield and the North-west.[5] The Sheffield & Midland Junction Railway set up its Provisional Committee during June 1836 to connect Sheffield with the Midland Counties line.[6] Its prospectus was issued during the third week of July,[7] detailing a Committee of

33 people, many of whom were amongst the Midland Counties subscribers. The Company's capital was set at £900,000 in £50 shares for the construction of a line linking Sheffield and the intended railway thence to Manchester with the Midland Counties Railway at its crossing of the river Trent and having a branch to meet the Mansfield & Pinxton Railway, so covering that portion of the original scheme which had been dropped from the Midland Counties Bill.

A meeting in Sheffield on 24 October 1836 received a report by Messrs Leather and Locke that there would be no difficulty in linking Sheffield with the North Midland Railway at Woodhouse Mill at a cost of £150,000. However, no line could be found between Woodhouse Mill and Clay Cross which would not run parallel with the North Midland line, and such a project would be unlikely to meet with Parliamentary approval. The line to Woodhouse Mill only therefore would be made, the Midland Counties Company, to whom these plans were presented, being invited to extend its own line to Clay Cross.[8] A further meeting was called for 31 October before the scheme faltered. Encouraged to believe that it would, nevertheless, be revived, on 11 November 1836 the Midland Counties Board decided to make immediate application to Parliament for powers to construct an extension from the authorised lines at Cranfleet to meet the North Midland Railway at Clay Cross, for which purpose Vignoles drew up a new line along the eastern side of the river Erewash, this time complete with several colliery branches.[9]

These actions of the Directors were confirmed by a Special General Meeting of the proprietors on 14 February 1837. The plans were naturally opposed by the North Midland Company, which regarded the project as directly competitive with its own line. Petitions against the Bill were also presented by the burgesses of Derby, complaining once again that their town would no longer be on the main trunk route. In the event, the Bill was found not to comply with Standing Orders,[10] and so it fell. The quarrel was then settled by the two companies agreeing to exchange at Derby all traffic to and from Rugby and beyond. The Birmingham & Derby Junction Railway Company, which had also been authorised in 1836 to build a line between Derby and Birmingham, together with the *Stonebridge Railway,* a competitive branch into the London & Birmingham Railway at Hampton-in-Arden, saw in this agreement an attempt to prevent it from obtaining any of the London traffic, and took the opportunity during opposition to a further Midland Counties Bill to have the arrangement declared void. The three companies were thus bound under the Midland Counties Railway Act of 10 August 1840 not to enter into any further restrictive agreements with one another.[11]

Another clash with the Birmingham & Derby Junction Railway occurred in 1837 over that Company's proposal to build a line from Tamworth to Rugby as an extension of the Manchester & South Union Railway, a directly competitive move not likely to commend itself to the Midland Counties Railway. That Company therefore joined forces with the London & Birmingham Railway and managed to get the Manchester line stopped short at Stone, while the Tamworth Bill was rejected as having failed to comply with Standing Orders.[12] Attempts were made in 1839 to revive the Manchester & South Union Railway as a direct line between Stone and Rugby,[13] when the Midland Counties Directors once again opposed the project, supporting instead the Churnet Valley Railway. This was designed to connect Manchester with Derby through Macclesfield, so giving an alternative route from the North-west to London, and several of the Midland Counties Directors were elected to its Provisional Committee.[14]

BUILDING THE LINE

Acquisition of the lands required by the Company led to several disputes. Due to their misunderstanding the Company's exact requirements, Leicester Council fixed the price of its land at £12,061 in the first instance.[15] When the Company's intentions were properly explained, this asking price was reduced to £9,250 against the Company's offer of £7,500, later increased to £7,750, a figure which the Council finally accepted.[16] On several occasions it was necessary to refer compensation claims to a jury, as in one action before Sheriff Swann at Nottingham. The plaintiffs claimed £6,000–£12,000, pleading development value against the Company's offer of only £450, which was based upon the consideration that the land was merely pasture on which grazing rights would not be relinquished. In this case the jury awarded £1,800,[17] but further powers had to be invoked to settle the value of the disputed common land in the Nottingham Meadows. The Company's Act of 4 July 1838 stipulated that this should be fixed by a jury drawn from the County of Nottingham, granted an additional year for the outstanding purchase negotiations, and set an extended time limit of eight years upon the completion of the works.[18] This Act also allowed the borrowing powers to be invoked as soon as half of the authorised capital (i.e. £500,000) had been paid up, and it also authorised a mineral branch to be worked by horses only at Mountsorrel – a development which was not in fact carried into execution until the Midland Railway Company had acquired fresh powers in 1859.[19]

The building of the line was divided into fourteen contracts, the first one to be let being No.2, for 6 miles 2 chains 10 yards of line between the west side of Carrington Street in Nottingham and a point near Long Eaton. It was secured on 22 May 1837 by John Taylor, Thomas Johnson, and Henry Sharp of Long Eaton at a price of £35,236, to be completed within one year.[20] Although all of the contracts had been let by the end of 1837, work was at first concentrated upon the section between Nottingham and Derby, with 400 men reported working vigorously along it in September.[21] William Mackenzie of Leyland in Lancashire had secured contracts 1, 3, and

5 (Northern Section) for 19 miles 3 chains 17 yards of line between Derby and Loughborough, including all three curves at Sawley and Long Eaton and the erection of the bridge over the river Trent, for £166,678 11s,[22] the Butterley Company supplying the actual ironwork under contract No. 4.[23] In its original form this bridge consisted of three cast iron arches, each spanning 100 feet and set upon stone piers and abutments. It was replaced by the present structure after the additional goods lines had been driven through Redhill in 1900, quarters of the Midland Counties Company's armorial device, incorporated as decorative features over the piers of the original bridge, being transferred to the parapet of the new one.[24] Between Loughborough and Syston, a distance of 6 miles 7 furlongs 6 chains 8 yards, contracts 5 (Southern Section) and 6 had been awarded to Edward Eckersley and William Worswick of Wigan, for £56,619,[25] and between Syston and Leicester contracts 7 and 8 were eventually re-let to Gordon and Hector Mcleod for £59,795 3s 8d.[26]

At the Annual General Meeting in 1837 it was decided, bearing in mind the earliest date fixed by the Act, to push on as soon as possible with the remainder of the line between Leicester and Rugby, and the whole of contracts 9 to 14 were awarded to David McIntosh for £250,629 10s 6d.[27] When it came to sealing the contracts, McIntosh at first refused to sign, or to offer any securities, wanting several changes in the specification. Responsibility for the works being divided at the Trent crossing between two committees, the one governing the southern portions therefore called a meeting of the main Acting Committee, which gave further instructions regarding the disputed terms. Since this move still did not produce the Contractor's signature, his securities, or any start upon the works, the company's solicitors, Berridge, Berridge & Macaulay of Leicester, who had been conducting the negotiations, were instructed to issue a writ in the Court of Common Pleas, and to consult Vignoles regarding the possibility of re-letting these

7. Constructed by the Butterley Company, the bridge across the river Trent south of Long Eaton, together with the tunnel through Redhill, also shown in Charles Lewsey's contemporary drawing, represented two of the major items of engineering on the line. This bridge was eventually superseded by the totally different structure now occupying the site. Derbyshire County Libraries.

contracts. A compromise was, however, reached upon the outstanding points regarding the supply of rails and certain other matters, the writ appears to have been withdrawn,[28] and work at last began in a field one mile south of Leicester.

The Directors had given some thought to the nature of the permanent way at their Board Meeting on 1 March 1837, when Vignoles had suggested laying the rails on longitudinal bearers. The London & Birmingham Railway Company was therefore approached for permission to lay down an experimental stretch of track upon its line at a cost of £500.[29] This was agreed and 534 yards were so treated near Watford, although the Midland Counties Company later decided to lay only 2½ miles of double, or 5 miles of single track of this type, the remainder to be laid either upon stone blocks or wooden cross-sleepers, with parallel form rails weighing 77 lb per yard in iron chairs.[30] The final choice was wooden sleepers from Leicester to Rugby,

with bearings of less than 5 feet as necessary, Messrs Bagnalls of West Bromwich securing the contract to supply all rails at £12 10s per ton. "Some portion of the way towards the Rugby Junction" was, however, laid with Evans's dove-tailed bridge-rails weighing 57 lb to the yard on longitudinal kyanised memel timbers measuring 9 by 4½ ins.[31]

The year 1838 did not open very favourably for the Company, as most of the works were held up for about two months by severe frost, and further trouble arose on contracts 7 and 8, which had been let in the first instance to Messrs Hindmarsh and Meredith. Although the partnership between the two principals had been dissolved, the Company was prepared to let Meredith continue alone if he could find the necessary securities, and to allow him until 1 October 1839 to finish the earthworks. It seems that he was unable to meet these terms,[32] and the contracts were re-let in March 1838 to Gordon and Hector McLeod.[33] These new partners seem to have been most dilatory, and by September 1838 it was reported to the Committee of Works that operations would have to be considerably speeded up in order to have the works completed within the time limit. Further warnings were given to these Contractors during March 1839, and in June they were warned that the works would be taken over by the Company unless more vigour was displayed.[34]

Satisfactory progress was being made elsewhere, but a ganger or *gaffer* named David Jones, who was working near Kegworth and had quartered some of his men upon the inhabitants, decamped with the gang's money and left his hosts wishing that they had not listened to what the newspapers called ''his steam promises''.[35] A particularly interesting episode occurred when Mackenzie was able to save paying out compensation for stoppage of the Derby Canal, at a penalty of £2 per hour, where it was necessary to divert the waterway near Spondon, by taking advantage of a normal repair stoppage. Gathering together all available labour from other parts of the line, he was able to complete the diversion during the time taken for the canal repairs, much to the astonishment of the local inhabitants. The employment of 300 men in such a small space must certainly have presented ''a very animated spectacle''. In order to induce the extra effort needed from his men, Mackenzie supplied daily allowances of beef and ale, to be consumed on the site, in addition to their normal wages.[36]

The contracts for laying down the permanent way on the Nottingham-Derby line were let on 8 May and 5 June 1838 to those contractors who had been responsible for the earthworks, No.1 at 5s 0½d per lineal yard and No.2 at 5s 6d, both sections on stone blocks and to be maintained and upheld for one year.[37] Laying of the blocks began at Chilwell on 13 August 1838, before quite a large audience.[38] Erection of the Trent bridge had also begun, and a start had been made on the tunnel through Redhill,[39] while 1,000 cubic yards of soil were being removed weekly from Knighton Tunnel. The embankment near to, and the bridge over, Humberstone Road, Leicester, were nearing completion.[40] The foundation stone of Nottingham station was laid in June 1838,[41] when it was also reported that over 3,000 men were at work between the river Trent and Rugby. Contracts for the laying of the permanent way on this portion were let to the section contractors on 25 January 1839 at 6s 9d for blocks or 3s 9d for sleepers on sections 3 and 5 (North), 7s 0d or 4s 0d on sections 5 (South) and 6, and 6s 6d or 4s 6d on sections 7 and 18. When it came to sections 9–14 the Committee refused to accept the proposal made by McIntosh, who said he would be unable to complete the task before 1 July 1840 and did not wish to maintain the track for the one year specified. His tender was 6s 8d for laying blocks and 4s 9d for sleepers.[42]

By the end of February 1839 most of the work on the Nottingham-Derby line was nearing completion, although there was still a considerable amount to be done near Chaddesden, particularly on the embankment there. The Contractor said that, providing the weather held firm, he could see no reason why this section should not be ready for opening by 1 May.[43] Woodhouse reported that

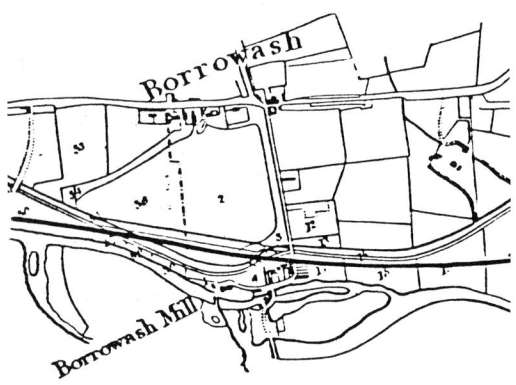

8. Plans for the diversion of the canal at Borrowash involved the construction of a new lock. The site of Mackenzie's penalty saving campaign of intense activity during a routine repairs stoppage of the canal lay farther west. Nottinghamshire Records Office.

9. The ornamental tunnel portal at the north side of Redhill, built to assuage opposition from the landowner, was faithfully mirrored when the goods lines tunnel was made alongside at the end of the century. R. Allen's *Nottingham and Derby Railway Companion.*

as many men as could usefully be employed were working on contract No.1, and two shifts were being operated in the tunnel at Redhill. When this was pierced on 9 February the event was appropriately celebrated at the Navigation Inn, Trentlock.[44] At the Rugby end of the line work was proceeding on the Avon viaduct, but the first arch collapsed when the timber framing was removed and it had to be rebuilt.[45] During the excavation of the cutting at Thurmaston, near Leicester, the workmen discovered a bed of plaster within five feet of the surface, and it was anticipated that this might bring additional traffic onto the line.[46]

By January 1839 a sub-committee had been set up to contract for the supply and repair of locomotives, of which 25 were judged to be sufficient, and for such other items of rolling stock as carriages, horse-boxes, and luggage wagons.[47] Delivery of the engines began by April, but the first one from Messrs Stark & Fulton had an unfavourable reception, Kearsley, the Locomotive Engineer, reporting that it would take about three weeks

to prepare it for traffic and attributing its condition to the manufacturer's inexperience.[48] As a result of this, Kearsley was requested to report upon the condition of each engine upon delivery from any of the suppliers. He criticised unfavourably another engine from the Butterley Company, stating that it lacked power, rolled heavily, and was an extremely unsteady runner, it having left the rails during trials.[49] A few of the Directors then began to venture out on trial runs from Nottingham to Long Eaton and back, one such journey being completed in only 16 minues on the outward journey and 17 minutes on the return, and being witnessed by large crowds at the lineside.[50]

OPENING BETWEEN NOTTINGHAM AND DERBY

Although it had been hoped to open the line between Nottingham and Derby on 1 May 1839, contract No.1 was not quite finished, and not until Thursday, 29 May, was everything ready to accommodate the official party of

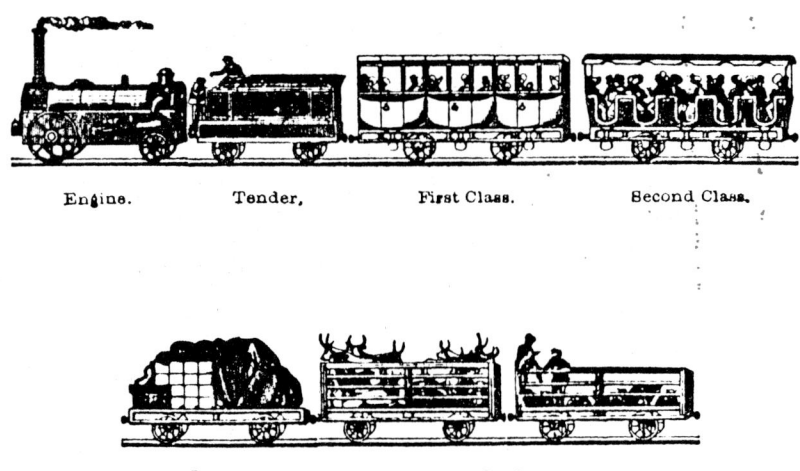

Engine.   Tender,   First Class.   Second Class.

Luggage.   Cattle.

10. A variety of carriages and waggons illustrated in R. Allen's *Nottingham and Derby Railway Companion.*

Directors and shareholders, with their friends. The Committee which had been responsible for the works north of the Trent had, at the Directors' request, made the necesssary arrangement for this official ceremony, instructions having been issued for Woodhouse to use his own discretion in regard to the temporary arrangements necessary at Derby.[51]

As early as 2 February 1836 the Town Council of Derby had considered the question of a single station for all three companies whose Bills were then before Parliament. Each had decided upon a separate station on *The Holmes,* which, had they been built, would have produced some complicated flat crossings on the approaches. The Council was at that time of the opinion that the Holmes would in fact be suitable for a combined terminal,[52] but further consideration showed that this site was likely to prove too restricted, bounded as it was by the river Derwent on one side and the Derby Canal on the other. In view also of the high costs likely to be involved at the Holmes it was later decided that Castle Fields would prove a much better site.[53] Even then it appears that the Midland Counties Company still intended to build a

separate station alongside that of the other parties. However, Woodhouse pointed out that £14,000 might be saved by joining forces, and this procedure was therefore adopted early in 1838,[54] the necessary arrangements being confirmed by the North Midland Railway Act of 1 July 1839 which empowered that Company and the Midland Counties Railway to purchase land from the Birmingham & Derby Junction Railway for the site and left the three companies to arrange between themselves as to the use of and the division of the cost of a single station.[55] It was later agreed that the premises should belong entirely to the North Midland Company, the others each paying six per cent on the money used to provide their own accommodation. Separate engine sheds and workshops were erected by each of the Companies on land between the station and the Derby Canal. Until the North Midland premises were ready, however, the Midland Counties Railway made use of a temporary station with no roof and only a wooden platform,[56] but neither its actual site nor the date of its demise is known.

At Nottingham it had originally been intended to take the line as far as the London

11. Complemented by the neighbouring Midland Hotel, not then in railway ownership, Derby Station, constructed by the North Midland Company as accommodation for the three railways meeting in the town, was eulogised in the pages of the *Illustrated London News* on 15 July 1843. Nottinghamshire County Libraries.

12. Having every appearance of the modern "open day" about it, with sightseers wandering about across the lines, this view of the interior of Derby Station during the visit by the Royal Agricultural Society was obviously derived from A.B. Johnson's unpeopled illustration at p.34 of Allen's *Midland Counties Railway Companion*. It indicates the peculiar layout with a single bayed platform at the right, behind the offices and booking hall. *Illustrated London News,* courtesy Nottinghamshire County Libraries.

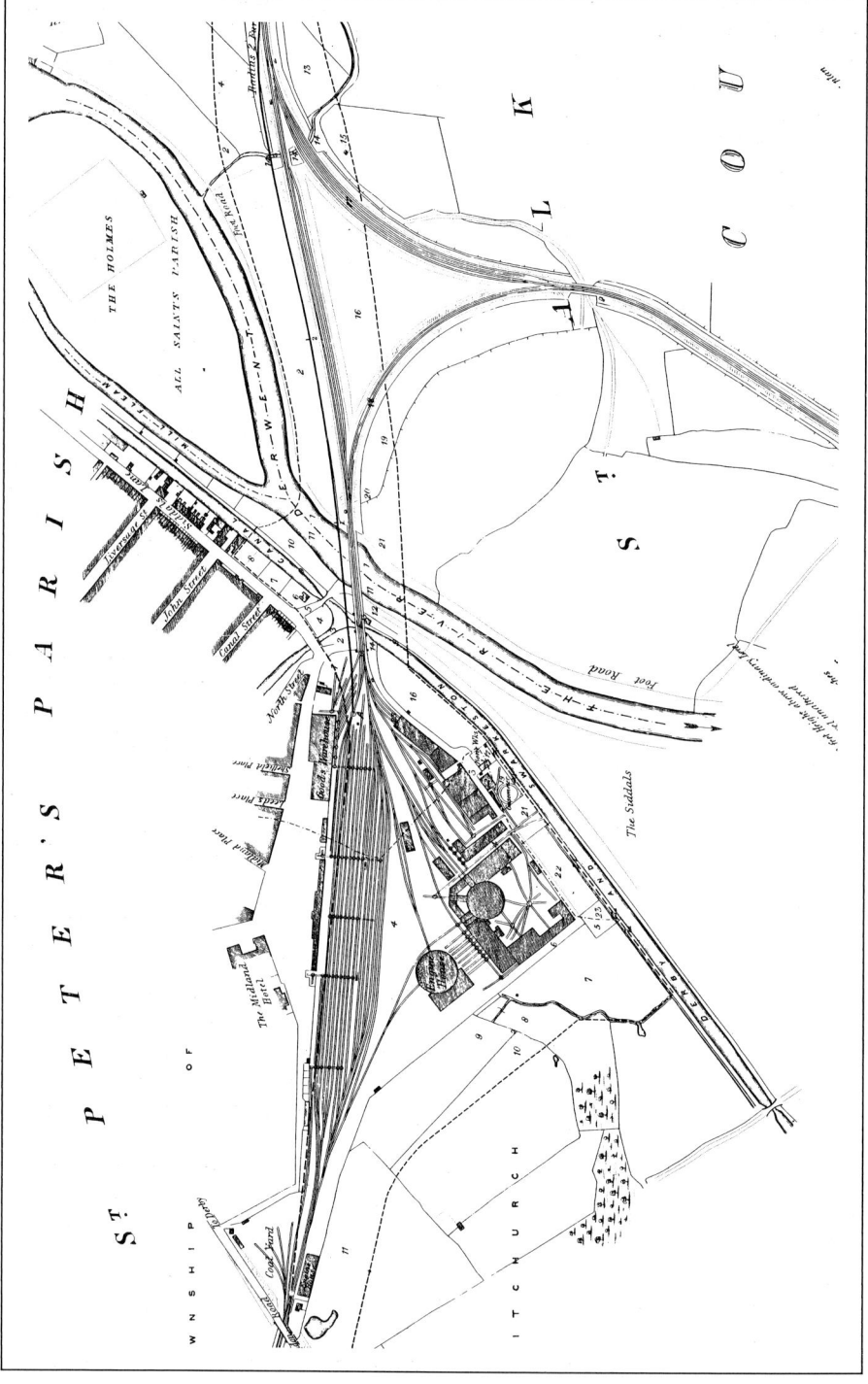

13. Site plan of Derby Station, as constructed

Parliamentary plans, submitted by the Midland Railway at the end of 1847 for making the goods depot at St. Mary's Bridge and widening the northern approaches across the canal into Derby Station, give a clear picture of the expanding layout. The single, bayed platform, used to accommodate the trains of the three original companies, may be particularly noted as well as the large number of sidings and the various locomotive depots and workshops established along the canal bank. Nottinghamshire Records Office.

Road, but the Act stipulated that it must not be extended beyond the western side of Carrington Street as projected from its existing junction with Leenside (now Canal Street). Access to the London Road would be along a new road to be laid down upon Corporation property and to be removed if it ever became redundant. A terminal was therefore built facing the end of this *Station Street* in the West Croft, with a Grecian Doric facade, 90 feet in length and built in stone brought from Darley Dale in Derbyshire. Accommodation was provided for an entrance hall, ladies' and gentlemen's waiting rooms, booking offices, and a residence for the Chief Clerk. There was a double shed for the trains, the north side serving arrivals and the south side departures.[57] According to Whishaw this was covered by a light iron roof, supported on the departure side by a brick wall with eight windows and on the arrival side, and along the middle, by two rows of nine cast-iron columns. There were four sets of metals and as many turntables outside the shed, with an intersecting cross-line communicating with a carriage wharf on the arrival side. There was also a through double road brick loco-motive *engine-house,* also on the arrival side of the line but a little removed from the passenger shed. Designed to hold up to six four-wheeled engines and four tenders, it measured 25 feet 8 inches wide, and the door openings at both end offered a clearance of 9 feet 8½ inches in width. The engine turntables were of wood and measured 14 feet across. Arriving passengers left the station by gates at the side of the offices opening into the road in front of the station.[58]

In order to provide a more direct access between this station and the Market Square, Nottingham Corporation agreed to extend Carrington Street by building a bridge over the canal, appointing a Committee for this purpose on 21 October 1839. The Railway Company's offer of £3,000 towards the cost was made conditional upon the provision of a bridge 50 feet wide and being allowed to retain the level crossing in Wilford Road instead of substituting a bridge there as the

Midland Railway afterwards did. Although the Corporation objected both to the excessive width of the proposed bridge and to the continued existence of the level crossing, it was made quite clear that the Railway Company would contribute only upon those terms, and agreement was eventually reached upon such lines. Tenders were invited, and the contract was won by Henry Sharp of Beeston on 23 July 1841 and finally sealed on 21 October for £4,814 plus the approach roads at a further £818.[59] Until all of these roads were ready, therefore, it seems that access to the station must have been gained over the canal bridge in Wilford Road. According to Allen's *Guide* "The road from the Station to the town is over a wooden bridge erected across the canal about a hundred yards from the Station house door; or there is another way by walking through the yard and omnibus gate and crossing a brick bridge called the New Bridge."[60]

The invited company assembled for the official opening of the line at noon in fine weather in the yard at the rear of Nottingham Station, special tickets having been issued bearing the arms of the four counties served, Nottinghamshire, Derbyshire, Leicestershire and Warwickshire. The number on each ticket corresponded with that on a reserved seat in the trains, the first three of which consisted of six carriages each, while the fourth one had only two. Each carriage bore the painted arms of the four counties and a flag atop and *Ariel, Sampson, and Hawk,* three of the company's own locomotives were standing ready, together with *Mersey,* an engine made by Galloway & Co of Manchester for one of the contractors. With their own flags and new uniforms, the police force added colour to the occasion, while the band of the 5th Dragoon Guards played martial airs and started each train away with the national anthem. The first one, consisting of four first and two second class carriages, left at 12.30 pm behind the locomotive *Sunbeam.* The second, which was hauled by *Ariel,* followed about ten minutes later, stopped for water at Breaston at 12.57 pm, and reached

14. This prospect of Nottingham Station, with and without the rainbow, was engraved for and published by R. Allen of Long Row, Nottingham in connection with his railway guidebooks. A surviving pencil sketch, including the rainbow but with slightly more bucolic characters in the foreground, upon which Allen's engravings may have been based, is illustrated here, courtesy of the Brewhouse Yard Museum.

15. Close-up views of the rear of the station are difficult to find, this one deriving from Henry Burn's coloured *View of the Town,* published in March 1846. Brewhouse Yard Museum.

16. Nottingham Station platforms

This view, taken from *The Illustrated London News,* 9 December 1843, shows the crowds assembled in the meadows at the rear of the station to witness the arrival of the Queen. It illustrates, better than any other view, the staggered arrangement of the platforms within the train shed. Nottinghamshire County Libraries.

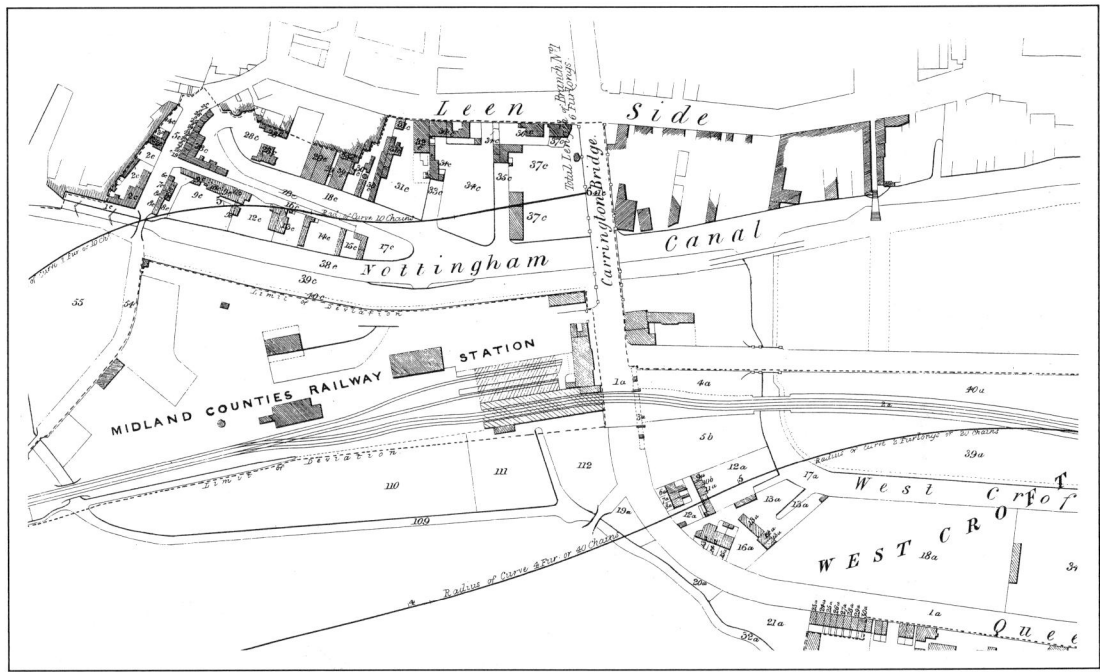

17. Site plan of Nottingham Station, as constructed

Parliamentary plans, lodged by the ''Ambergate'' company towards the end of 1846 for making alternative station arrangements in Nottingham, show in some detail the layout of the original Midland Counties station *after* it had been adapted to cater for the Lincoln trains. The curious disposition of the arrival and departure platforms in the original train shed, both on the north side of the relevant track, and the total absence of the sidings and turnplates mentioned by Whishaw may be noted. As this plan was made some time prior to the commissioning of the new premises in Station Street it is unlikely that it truly represents the full extent of the track layout, especially around the warehouse also served from the canal. Nottinghamshire Records Office.

18. Opening Ceremony

A Parker's drawing, engraved by G. Hawkins Junr and printed by Day & Haghe, purports to show the opening ceremony, with the train emerging at a very acute angle from the station. Nottinghamshire County Libraries.

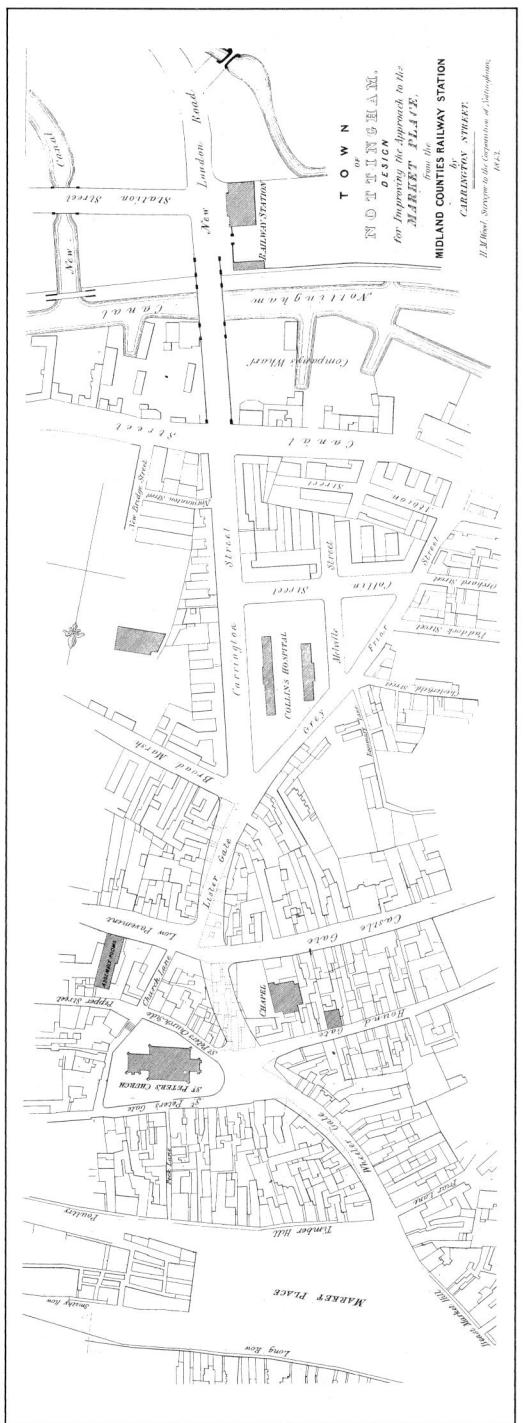

19. A plan for improving the approach to Nottingham Market Place from the Station, by widening and connecting together Wheeler Gate and Lister Gate, compiled by H.M. Wood, Surveyor to the Corporation, 1843. The wide canal bridge in Carrington Street would seem to have been completed before this date. Brewhouse Yard Museum.

20. Another detail from Henry Burn's *View of the Town* shows a four-coach train passing the foot of Nottingham Castle Rock. Brewhouse Yard Museum.

21. Trains were short, as this view of the approach to Nottingham through the Lenton Meadows, published by R. and F. Allen, indicates. The representation of ballast-topped and widely spaced cross-sleepered track may be noted. Brewhouse Yard Museum.

Derby at 1.19 pm, having taken 44 minutes for the journey. Great crowds gathered all along the line, and particularly at Beeston, while the men who had worked on the line were assembled at another spot to give a very lusty cheer.

Some of the passengers on these trains had gone to stay in Derby and others to visit various parts of the town in the short time available to them, but most returned to the station by the time appointed for the return journey. It was at 2.31 pm that *Ariel* and her train started back to Nottingham, passing very slowly over the canal bridge. Then, as steam was laid on, away went the whole assembly at the tremendous pace of over 30 mph, Nottingham being reached at 3.13 pm in 42 minutes, cutting the outward journey time by two clear minutes. Reference to the slow start across the canal in Derby could indicate that the trains had been on North Midland Company territory before crossing the Swarkestone branch of the canal to curve sharply eastwards over the Little Eaton arm on the way to Chaddesden and Spondon. Between Spondon and Borrowash the line ran narrowly between the river Derwent and the

23. Another view of Beeston Station, from the lineside. The waiting shelter, as well as the station building and the road, seem to be extremely well protected against boarding passengers. R. Allen's *Nottingham and Derby Railway Companion.*

Derby Canal, a stone retaining wall supporting the latter at a slightly higher level. Beyond Draycott and Sawley the Birmingham-Nottingham road was crossed by an overbridge to reach the junctions with the works on the Leicester section, Nottingham being entered through Attenborough, Beeston and Lenton, where there was a bridge over the Nottingham Canal before the line entered the Kings Meadows below the Castle Rock.[61] The trains were played in to the strains of *See the Conquering Hero Comes* as the passengers alighted. When everyone had been reassembled, they were treated to the usual cold collation at the Company's expense, the gentlemen facing the ladies across the tables. Speeches and toasts to Her Majesty, the directors, the contractors, and everyone else involved then followed.[62]

Public services began on Wednesday, 4 June 1839,[63] with four trains each way on weekdays and two on Sundays. Protests against this desecration of the Sabbath were quickly forthcoming, particularly after a special train was run to take sightseers to Beeston Wakes.[64] The Directors replied that this particular instance had been unauthorised and would not be repeated, but at a

22. The present buildings at Beeston Station date primarily from 1847. Prior to their erection the community was served by a much smaller single-storey booking office alongside the level crossing. R. Allen's *Nottingham and Derby Railway Companion.*

24. Drawn before the line was doubled, the original station at Long Eaton occupied a site alongside Meadow Lane crossing. R. Allen's *Nottingham and Derby Railway Companion.*

25. Site plan at Long Eaton Junction

In its Parliamentary plans of November 1847, the Midland Railway Company sought powers to build a tighter curve connecting the lines diverging from the riverside junction at Trent. Although this line was not actually made until 1862, and then under powers obtained under another Bill, the original plans give quite a detailed view of the layout around the Trent triangle. Long Eaton station is shown alongside the Meadow Lane crossing, together with the lines of the Erewash Valley Railway, recently completed and intersecting the Nottingham-Derby line on the level at Platt's crossing. Nottinghamshire Records Office.

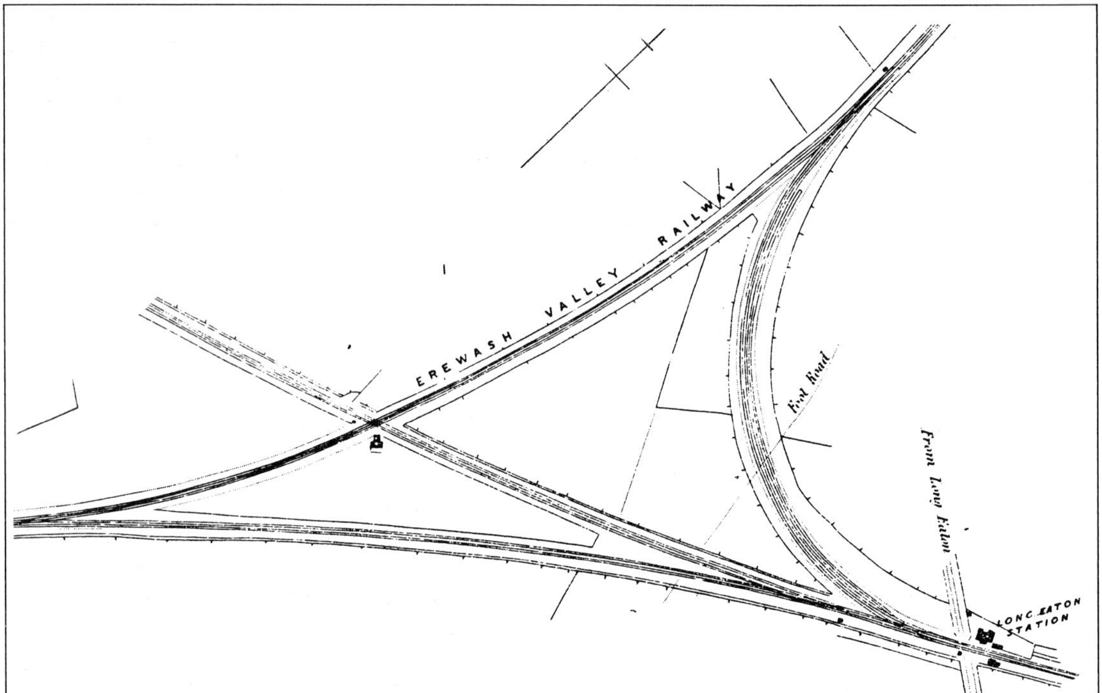

subsequent meeting of the Board a proposal that only those trains required by Parliament should be run on Sundays was rejected in favour of a counter proposal that the matter be left to the discretion of the Directors.[65] The line, single at first, was soon doubled between Derby and the junction at Sawley, in preparation for the opening of the remaining sections to Leicester and Rugby.[66] Intermediate stations were provided at Beeston, Long Eaton, Breaston and Borrowash only at first,[67] Spondon being added on 11 November 1839, as mentioned in Allen's *Guide* to the line. Within twelve months the name of Breaston Station was changed to *Sawley,* in order to avoid any further confusion with the one at *Beeston.* This station was situated beside the level crossing on the road between Breaston and Sawley, not on the main road at Sawley Junction, where the station was not opened until 3 December 1888, surviving down to the present day as the last of five *Long Eaton* Stations.

The earliest station in Long Eaton was that constructed by the Midland Counties Railway on the east side of the crossing in Meadow Lane, which served also to cater for trains off the Erewash Valley line when that was eventually opened (at first only as far as Codnor Park) in 1847. Another station, referred to as *Toton for Long Eaton,* was built for the Erewash Valley Railway on the north side of the crossing in Nottingham Road, being moved the short distance southwards to the crossing in Station Road in 1863. The year before that alteration witnessed the replacement of the Midland Counties station by the new establishment at *Trent Junction,* coincident with the opening for passenger traffic of the extension of the Erewash Valley line through Alfreton to Clay Cross,[68] the source of so much anguish to the projectors of the Midland Counties Railway. Draycott and Attenborough Stations were not opened until May 1852 and 1 September 1864 respectively, and the original station at the crossing of the Breaston to Sawley road was closed to traffic on 1 December 1930,[69] all of these events occurring long after the demise of

the Midland Counties Railway as a separate company. They are recorded here in order to clarify the situation which obtained in 1839, shorn of these later developments.

COMPLETION TO LEICESTER AND RUGBY

In the meantime, work was proceeding swiftly on the main line south of the river Trent, 605,793 cubic yards of earth having been moved by 4,499 men, 463 horses, 4 locomotives, and a stationary engine between 18 May and 13 July 1839.[70] September saw the signing of the contracts for building the intermediate stations, the costs of which varied considerably, from £904 for Loughborough and £476 for Kegworth to £386 for Ullesthorpe. A gatehouse at Countesthorpe was erected for £325.[71]

It had originally been planned to build the Leicester Station between Rutland Street and Yeoman Street, with two branches forming the connections into the main line. On 3 November 1836, however, the Committee responsible for this section of the works decided that a site either at the top of Northampton Street or at the foot of Campbell Street would be more convenient. Accordingly, it was decided that powers should be sought to abandon the originally authorised branches and to substitute a modified arrangement, although at first on similar lines.[72] Since negotiations for the purchase of the land required were carried through without opposition, on 17 January 1837 the Committee decided that it was unnecessary to continue with the application to Parliament, which had already been instituted,[73] and a through station was built on the main line at the foot of Campbell Street instead.

Plans for the building were drawn up and the contract for erecting the whole premises, including engine sheds, was let on 17 July 1839 to William Waterfield and Thomas Smith, Leicester builders, for £12,487. On 20 July, however, the contractors reported that they had made an error of £2,317 in their tender. This resulted in the tenders being reopened,

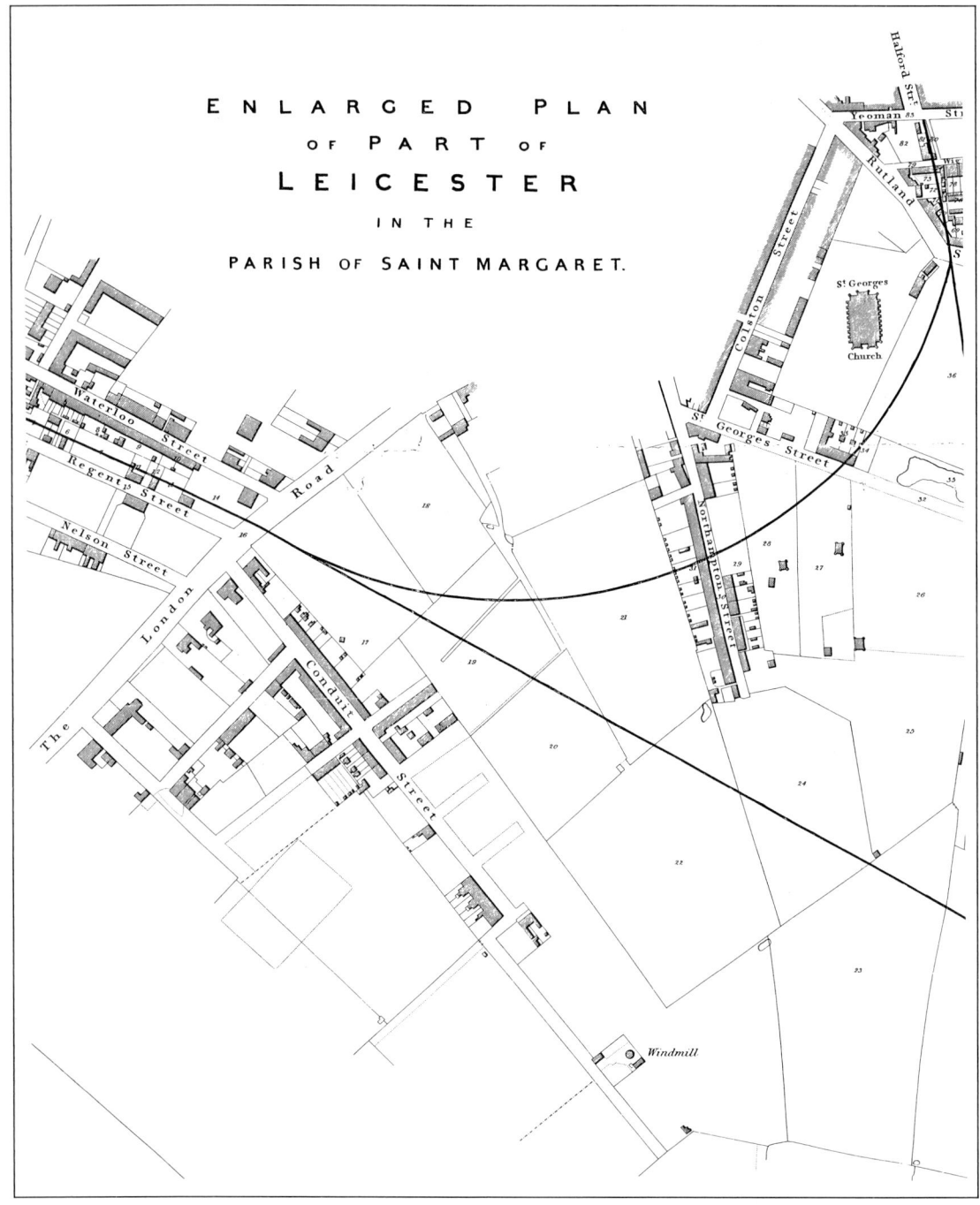

26. Various alternative sites for a terminus in Leicester off the main line were considered before it was finally decided to build a through station approached via Campbell Street. Nottingham Records Office.

27. At Leicester the Midland Counties Railway built its station in Campbell Street, well illustrated in this view of the decorations welcoming the Queen during her progress back to Windsor after her visit to Belvoir. *Illustrated London News,* courtesy Nottinghamshire County Libraries.

but Messrs Waterfield and Smith once again secured the contract, this time for £14,804, on 27 July.[74] The booking office, which was approached through Campbell Street, was divided by an elegant iron railing separating the first class passengers from their poorer companions. The Company's Offices and Board Room occupied the upper floor, above the booking offices, ladies' and gentlemen's waiting rooms, refreshment rooms, etc., being lit throughout by gas.[75] There was only one platform – 165 yds in length – which, by means of a loop line and connections at each end and in the centre, served both up and down trains after the manner which obtained at Cambridge until latter years.[76] With certain modifications, indicated on later plans, these premises remained in use until the completion of an extensive rebuilding on 12 June 1892.[77]

In addition to the troubles already mentioned over contracts 7 and 8, the Directors began to worry about McIntosh's slackness on the work between Leicester and Rugby. On 19 November 1839 he asked that the public

28. The same train of specially decorated carriages and luggage waggons was used throughout the Queen's progress on the railway. It is shown here departing from Leicester. *Illustrated London News,* courtesy Nottinghamshire County Libraries.

opening of this section should be postponed until 1 July 1840 – as it virtually was – but the Committee refused to agree at this time.[78] It should be pointed out in all fairness that this was the most difficult part of the works, as the line had to cross very hilly country, with heavy earthworks, deep cuttings and two substantial viaducts. On 18 February 1840 Woodhouse informed the Board that 700 tons more rails than were already under contract would be needed. As the price of iron had fallen by £1 per ton, Bagnall's were invited to supply the amount required, providing that they would accept the contract at £11 12s per ton delivered as and when required.[79]

In preparation for the extended traffic requirements, further staff appointments were made and more coaches, including closed ones for the night services and luggage vans, were ordered at the Board Meeting on 18 February 1840.[80] It had been estimated that some 20 tons of coke would be required daily when the line was fully open and, after experiments which showed that the quality produced in the Company's own ovens at Riddings was superior to that obtained from Poynton in Cheshire, sufficient ovens were ordered to be added to produce the necessary amount.[81] On 10 April, however, the Board accepted Mr Silverwood's offer to supply coke for three years from 1 July 1840 at 17s per ton for the first three months and 16s thereafter.[82] Although the Company ordered further advertisement for tenders on 14 April 1843,[83]

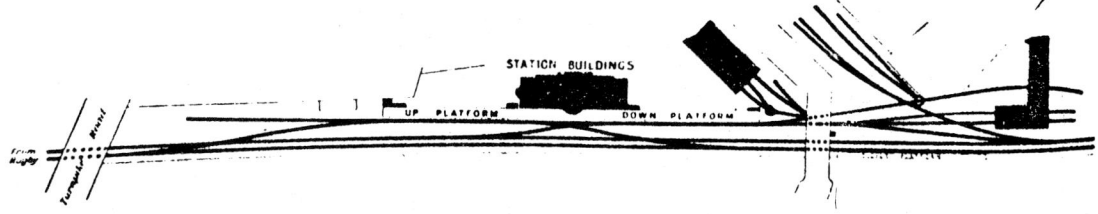

29. As built, Leicester station was provided with a single platform offset on the northbound or "down" side of the main through lines. Traffic was brought into opposite ends of this long and continuous platform through facing crossovers. The provenance of this particular illustration is not known. LMS Magazine.

the same vendor secured a further twelve months' contract from 1 October 1843 at 16s 9d delivered at Long Eaton Wharf, Silverwood to use and maintain the Company's own ovens at an annual rental of £200.[84]

Both the Nottingham–Derby and Derby–Leicester curves passed over the Erewash Canal at Sawley, on 26 foot span bridges which allowed a 20 foot waterway and 6 foot towing path.[85] A side-basin was thus very conveniently excavated alongside the Leicester line, served by an inclined siding and extensive coke store, built between the basin and the main line.[86] On 4 May 1840 a Mr Swanwick was appointed Clerk at the depot at £130 per annum plus house and coals all found,[87] and he was joined by a Mr Fletcher of Bath as Manager at £75 plus house on 29 June.[88] It was later ordered that a small forge was to be erected here at a cost of £25,[89] and the premises were gradually expanded to provide the Midland Railway Company and its successors with their extensive sheet and sack stores. Although the buildings have now been given up for railway purposes, many still survive together with the canal basin, given a new lease of life as a marina for pleasure boats.

Plans were made to accommodate four carriers at Leicester and two at Loughborough, and in March 1840 arrangements were made for the building of the necessary wharehouse accommodation.[90] About the same time, the Works Committee was invited by Dr Arnold, the noted Head-master of Rugby School, to bring the labourers working nearby to his Church services.[91] On 27 April the Secretary reported that he had good authority for believing that the North Midland Railway would be brought into use on 20 June and not, as previously expected, at the end of that month. Woodhouse explained that a single line would be ready by 18 June, but that it would take another fortnight to prepare the other track. In point of fact, the North Midland line was not brought into use until 1 July.[92] When the Board met on 4 May, however, the journey time between Derby and Leicester was fixed at 1½ hours, the drivers to be instructed that they would be fined for arriving *before* time.[93] Immediately after this meeting the Directors proceeded to inspect this section of line before bringing it into public use on the following day, 5 May 1840.[94] A service of four trains per day in each direction was introduced with no special ceremony.[95] On Sundays there were only two trains in each direction.

Work continued on the line between Leicester and Rugby, so that it was completed in time for a formal opening ceremony on 29 June.[96] The Secretary had in fact issued a public notice during the previous week,[97] stating that the line would be brought into general use on 1 July, along with the North Midland Railway, but local traffic began on Tuesday, 30 June 1840, making use of a temporary station probably situated at the northern end of the Avon viaduct alongside the Oxford Canal at Rugby.[98]

30. The carriages of this marshalling train, which were shown closely coupled in the original pencil sketch although not attached to the engine and luggage or brake vehicles, had been uncoupled and slightly separated by Allen's engraver. *Midland Counties Railway Companion,* courtesy Nottinghamshire Public Libraries.

The question of the station and junction with the London & Birmingham Railway at Rugby was first considered at a meeting of the Midland Counties Works Committee on 1 November 1837, when an approach was suggested regarding the two companies combining to build a new station at the junction of the two lines.[99] Although land purchase negotiations were begun in March 1838,[100] little progress seems to have been made towards finalising the matter until 22 May 1839, when the attention of the Acting Committee was drawn to the problem.[101] On 6 February 1840 representatives of the two Companies came to an agreement that the London & Birmingham Railway should relinquish its existing station in Leicester Road, short of the proposed junction, and use a new one at the junction itself, to be built at the expense of the Midland Counties Railway.

These premises would be leased to the London & Birmingham Railway at a rental of £100 for 99 years, the booking office and general management to rest in that Company's hands with an allowance to be made annually on account of work performed for the Midland Counties Railway.[102] Agreement was also reached that the London & Birmingham Company would provide separate engines to work the through mail trains from the Midland Counties line at the latter party's expense.[103]

At a further meeting on 13 March 1840 it was agreed that a connection should be made into the down line of the London & Birmingham Railway, that is a flat crossing over the up line so as to form a running connection into the Midland Counties line, in order to facilitate through working, and this must surely have been completed at the date of the opening of the Midland Counties Railway from Leicester, when through traffic was handled off the North Midland Railway at Derby. The wording of the London & Birmingham Railway minute, quoted by Dicey when reporting to his own Board, is curious. ''A crossing be made from the Down line of the London & Birmingham Railway … in addition to the Turnplates, to enable certain of the Midland Counties trains, to be mutually agreed upon, to pass onto the Midland Counties line without using such Turnplates. Should it at any time hereafter be found practically inconvenient, as respects the London & Birmingham traffic, for carriages to cross the London & Birmingham line on a level, the Midland Counties Company engage when called upon to make the Road marked upon the Plan for effecting the crossing by a Bridge which the Midland Counties will in that case construct; and they further engage, when called upon, to make a foot bridge for persons passing to and from the Booking Offices''.[104] The very idea of using turnplates to accommodate through carriages, one at a time, seems at this distance in time to be too ridiculous even to contemplate, especially when juxtaposed with the suggestion for what must have been the earliest flying junction ever formulated. The contract for the new station was not let until 23 October 1839, once again to Messrs Waterfield and Smith of Leicester at the price of £5,651.[105] Since it was necessary to interfere with the Oxford Canal Company's premises, agreement was reached as to an exchange of land and a charge of £800 upon the Midland Counties Railway towards moving the canal wharf,[106] but it is not known when any of these works were completed. In 1842 Whishaw described the new station as having several lines of way and a passenger shed of considerable length, the roof of which was supported at the front by 42 cast-iron columns and at the rear by a brick wall which also formed the front of the booking offices and waiting rooms. ''On either side of this shed, in which there is only one line of way, with side-spaces, each of 2½ feet, is a paved platform 9 feet

in width''. He gave an entirely different description of the London & Birmingham station with wooden platforms (one 8 feet 10 inches and the other, ''between the two ways'', only 2 feet 9 inches wide and 7 inches high above the rails) at the west side of an embankment 26 feet 5 inches wide.[107]

Apart from the bridge over the river Trent and the tunnel at Redhill, which have already been mentioned, the other major works

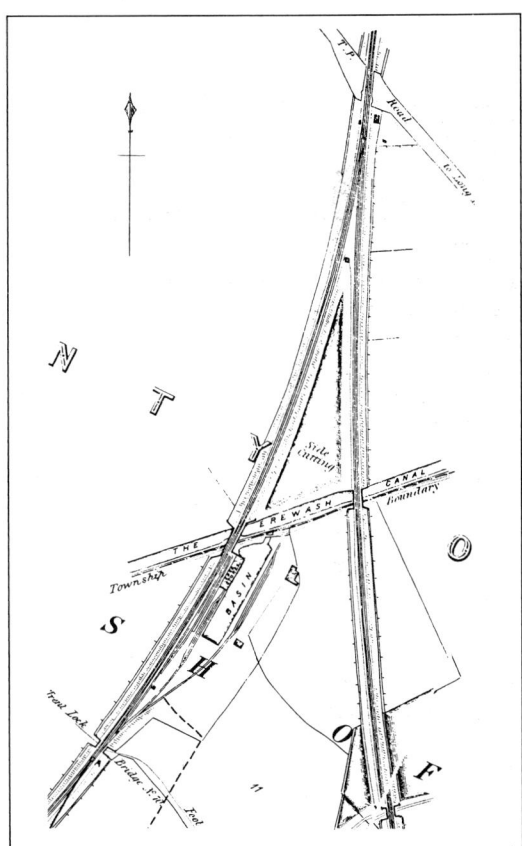

31. Another detail from the Midland Railway Parliamentary plans of November 1847, seeking sanction for the Trent south curve. This section shows the layout of the canal basin which formed the basis of the later Sheet Stores. Nottinghamshire Records Office.

between Long Eaton and Rugby included two further bridges over the river Soar, one of them comprising four arches at Normanton and the other just south of Loughborough at the junction with the artificial cut of the Leicester Navigation. Sileby village was virtually cut in two by a 40–50 foot high embankment, which continued across the river Wreake and the Fosse Way. The short tunnel under the New Walk in Leicester was followed by another, slightly longer one of 104 yards, under the Freemen's Common. About half a mile beyond Wigston, Crow Mills viaduct preceeded a series of extensive embankments and cuttings, which resulted in some peculiarly sited station buildings. At Broughton Astley, the platforms were reached by a flight of 15 steps up the embankment, while at Ullesthorpe the buildings were at the top of the cutting and served the Chairman's residence nearby. Beyond the level crossing over Watling Street there was a 66 yard tunnel at Gill's Corner into the valley of the River Swift, followed by a bridge over the Oxford Canal and the viaduct over the Lutterworth road and the river Avon.[108] Other intermediate stations were sited at Kegworth, Loughborough, Barrow, Sileby, Syston, and Wigston, but not all of the trains stopped at Barrow, Sileby, Wigston, Broughton Astley or, until early in 1841, at Kegworth.

The Midland Counties Railway cost some £30,019 per mile, comparing favourably with most of its contemporaries, the London & Birmingham Railway having cost £52,882, the Great Western Railway £56,372 and the North Midland Railway £45,871 – though the Birmingham and Derby Junction Railway had been built for only £24,683. Total expenditure had in fact been £1,725,693,[109] additional powers to raise or borrow the necessary extra amount having been obtained in supplementary Acts of 1840 and 1842.[110] An average of 94,112 cubic yards of earth was moved for each mile of the line, the total amount having been 5,335,000 cubic yards.[111]

4

Operating the Railway

The fares initially levied between Nottingham and Derby, amounting to 4s first and 2s 6d second class, were higher than the prevailing rates by road, at 3s inside and 2s outside the coaches, and the Company soon found it expedient to reduce them. From 23 June 1839 they became respectively 3s 6d and 2s, with intermediate stages in proportion.[1] Third Class accommodation was not available until the beginning of September, and the rate for this was soon increased from 1s to 1s 6d from 5 June 1840, when the Second Class fare once again returned to 2s 6d.[2] Small parcels travelling through between the termini were charged between 6d, for weights up to 14 lb, and 1s for those between 28 and 56 lb, heavier *Van Goods* items being rated at 1s 6d per hundredweight.[3]

From the start, traffic receipts were very good, never falling below the Parliamentary estimates. During the first six months an average of 2,500 passengers were carried each week, paying about £250 in fares, an additional £42 being realised on merchandise traffic. During Beeston Wakes fair in July 1839 a number of additional trains were put on at reduced fares, including 2s return from Derby and, when nearly 4,000 passengers were carried on the Sunday, the Company experienced a great deal of criticism.[4]

Between Nottingham and Leicester the fares were initially 6s and 4s 6d first and second class respectively, and between Derby and Leicester 6s and 5s.[5] Limited third class accommodation became available on 21 May. The weekly receipts now rose to nearly £600,

reaching close on £800 shortly before the completion of the line to Rugby. Of the five daily and four Sunday workings then provided in each direction, most made connections at Rugby for Euston and at Derby for Sheffield[6] and the North. One working in each direction continued through direct to and from Euston.

As already remarked, a limited amount of third class accommodation was provided on the early and late trains between Nottingham and Derby from 3 September 1839,[7] and from Leicester on 21 May 1840. A curious feature of the latter innovation was its advertisment between Leicester and Nottingham only, although it appears unlikely that the Derby portions of the trains, which seem to have been separated at Trent Junction, did not follow suit.[8] The accommodation provided was distinctly spartan, and designed to encourage full use of the superior categories, the third class vehicles first introduced between Nottingham and Derby being no more than open wagons with doors. There were no seats, their absence being considered small inconvenience on such a short journey, and the sides were no more than 3 feet 9 inches high. Third class passengers appear to have been booked as far as Rugby from the opening of the southern section of the line, although the London & Birmingham Railway did not yet carry them.[9] A curious passage in the *Notice of Opening* suggests that passengers would be booked through to Euston from each of the First Class Stations, i.e. those served by every train, with the unlikely exception of Leicester itself. Passengers to inter-mediate stations on the London & Birmingham Railway had to rebook at Rugby and possibly make their own way between the two stations there until the new one was completed.

Even with the possible inconvenience of this arrangement, the completion of the Midland Counties Railway provided a route between the North through Derby to London appreci-ably faster than that by the Birmingham & Derby Junction Railway, being shorter by 11 miles. It was to the eternal misfortune of the Midland Counties Railway that the rival route had already been in use since 12 August 1839, although the initial traffic between Leicester and Rugby was distinctly encourag-ing. During only the second week of operation the *Railway Times* remarked that the 9.30 am through train from Euston, scheduled to reach Nottingham and Derby in 5¾ hours, consisted of 12 carriages on the Tuesday, 17 on Wednesday, and 19 on Thursday.[10] Passenger receipts immediately doubled, reaching a maximum of just over £2,000 during one week in August, only to fall off with the onset of winter to a minimum of about £1,000 in December. During the first six months of full operation working expenses amounted to £32,119 16s 10½d, 55 per cent of the £52,221 10s 1½d receipts. No dividend was declared.

On 17 August 1840 the night mails began to be carried on the Midland Counties line instead of over the Birmingham & Derby Junction Railway as previously, the special trains carrying them being available to the public upon payment of a 2s surcharge.[11] The keenness of the management to cater for all traffic offering is shown by the report of a number of special trains held in readiness for the large crowds expected to converge upon Leicester one day during October. Such trains had already been put on in connection with exhibitions in both Nottingham and Leicester, and it was reported that on one such occasion a train which left Nottingham in the early morning and was strengthened with a Derby portion added at Trent Junction

before proceeding to Leicester at a fairly rapid speed, carried over 2,000 passengers. The combined train of 73 carriages was hauled by as many as four engines.[12] On another occasion a large number of passengers was conveyed between Derby and Nottingham in two special trains at fares of 3s first, 2s second, and 1s third class, with a free return journey.[13]

During its first two years of operations the Company made frequent alterations in the timetable, often of a very minor nature. The actual number of trains running between Nottingham and Derby was not altered until November 1839, when an early morning train was introduced which, from 6 April 1840, carried Nottingham mail from the Birmingham & Derby Junction line.[14] The service was thus increased to five return jour-neys and, soon afterwards, to three Sunday trains as well. High speeds may not have been demanded by the easy timetables, but were occasionally achieved, despite the threat of the fine upon the drivers for *early* arrivals. During 1840 a correspondent of *The Derby Mercury* remarked that his journey by the early morning train to Nottingham had often been completed in 22 minutes, and that speeds often exceeded this average of 42.6 mph with perfect safety. The scheduled time for this particular train was one hour![15]

At the end of August 1840, when the service between Leicester and Rugby was *reduced* by one train in each direction,[16] Spondon was being served by only two of the Nottingham and Derby trains, one each way.[17] By the end of the year it had obtained a much better service at the expense of Beeston. The withdrawn service to and from Rugby was restored, with London connections, from 7 September 1840,[18] and by this means it became possible to make a day visit to London, until both the up train and the early morning down train were withdrawn before the end of the year. Birmingham connections were maintained at Rugby by several of the trains.

Goods traffic began to pass through Rugby on 7 September 1840, after delays probably occasioned by the previous lack of exchange facilities there.[19] In October, third class carriages were added to the mid-day trains in connection with the third class trains introduced by the London & Birmingham Railway at the beginning of that month. When this practice was extended to the North Midland Railway from 8 March 1841, *The Preston Chronicle* was able to remark that it had become possible to travel between Manchester and London by the Manchester & Leeds Railway, Derby, and Leicester, at a fare twenty to thirty per cent cheaper than via the Grand Junction Railway, upon which this privilege had not yet appeared.[20]

The line to Rugby had been open for less than five months when, about mid-day on 24 November 1840, part of the Crow Mills viaduct, about 3½ miles south of Leicester, collapsed into the Leicestershire & Northamptonshire Union Canal.[21] Until a temporary timber bridge could be erected, trains were worked up to points on each side of the gap and passengers used flights of steps down to ground level in the course of their promenade between them, although it was reported that an omnibus was also in attendance to convey those passengers with luggage and the inspectors, etc. Certain traffic, including the night mail, was diverted over the Birmingham & Derby Junction Railway, under offers of assistance to convey urgent traffic,[22] until the timber structure was ready for use on 10 December.[23] The Midland Counties Railway held Mr Betts, for the Contractor, McIntosh, responsible for the restoration of the viaduct and intimated that, if this was completed as rapidly as possible, they would not press for compensation for loss of revenue.[24] As McIntosh objected to any such arrangement, the Company was forced to undertake the work on its own account, and a permanent replacement structure was not begun until after the winter. It fell again afterwards.

Similar troubles were also experienced on Belgrave embankment in Leicester, which had been weakened by the rains, one of the Sunday morning trains from Leicester to Derby being derailed here during November 1840. The driver was later punished for excessive speed, in view of the state of the road, his engine having turned right around with a certain amount of damage being done to the train. As a result of this accident an engineer was appointed to watch the state of the road, speed limits were fixed, and permanent signals were erected where caution seemed necessary.[25] Traffic was again interrupted in January 1841, when a rapid thaw caused a sewer in the New Walk, Leicester, to burst and throw down the wing of the bridge. The up mail train ran into the debris and was delayed for three hours, but there were no injuries. On the following night the river Soar overflowed near Loughborough, washing ballast from the blocks and making the line impassable. Passengers on the mail trains travelled by chaise between Loughborough and Leicester, but the track was restored for the next morning's trains.[26]

Frequent train alterations continued during 1841, sometimes in connection with altered services on the connecting railways. These included the withdrawal, in October, of the evening up slow train to Leicester[27] and, from 1 June, of the night mail trains to and from Nottingham. Connections were instead provided with the Birmingham & Derby Junction Railway mail trains at Derby which, begun on 5 April, were the direct forerunners of the celebrated *Tamworth Mails*. Until 1 May 1841 the mails for York and the North had been exchanged at Rugby but, following a request from Hudson on behalf of the North Midland Railway, they were henceforth carried on through trains.

Although the sprung suspension of other carriages had been improved towards the end of 1840, the third class vehicles were still very spartan, and it was only after a serious accident on the Great Western Railway, which led to representations from the Board of Trade, that the sides of these carriages were raised to the safer height of four and a half feet early in 1842.[28] There was still no roof, however, and the Management Committee's response to a letter from two ladies whose clothes had been damaged by cinders, was to express regret that they had not travelled by a better class of accommodation![29] Some of the early second class carriages were also open at the sides, being glazed only towards the end of 1842, when some at least of the third class carriages were also fitted with seats.[30] The returns of the Railway Department of the Board of Trade, published on 10 February 1845, indicate that the Midland Counties Railway must have retained some Third Class carriages without seats or roofs until the end of the Company's separate existence. According to Whishaw, however, writing in 1842, "Stanhopes", as these open carriages were termed, "were used on this line when first opened to the public, but have very properly been discontinued". Glazed coaches, possibly second class vehicles, were reported as working between Nottingham and Derby only in January 1843.[31]

COLLISIONS AND DERAILMENTS

From February 1843 it became the Company's regular practice to run an empty coach between the tender and the remainder of the train in the interests of safety, injuries being generally worse in the leading coach when trains collided or left the rails. Three months later some discussion centred upon the most suitable position for horse boxes, following a fire which began in some waste wrapped around the wheel of a gentlemen's carriage. Due to the placing of a horse box at the front of the train, it had been found very difficult to attract the driver's attention. During the subsequent enquiry, it was found that there was no commonly established practice amongst other companies, but it was agreed that guards would henceforth be provided with two portable fire buckets.[32] In June, enquiries were made into a Mr Petit's method of communication between the carriages and the engine of a moving train, and a box of Cowper's fog signals was ordered for trial purposes.[33]

During the first few months of operation, while the railway was still a topic of public interest, many minor accidents were reported in the local newspapers. The majority were either derailments caused by the unconsolidated state of the track, or cases of people and animals being run down whilst straying onto the line, the Board finding it necessary to issue a reminder of the penalties to which trespassers were liable. There were two cases of trains running away into the Nottingham terminus, the first, in September 1839, leaving one man severely injured and the second, in December 1842, causing some damage to the premises.[34] The generally poor state of the track in its early life is well brought out in the report of a derailment involving the 7.30 pm train from Leicester to Derby on 26 May 1840. About one mile south of Barrow the engine, tender, and a luggage van were buried in sand, and only the last vehicle of the train, a third class carriage, remained on the rails. Although a messenger was sent to Long Eaton for another engine, it was soon discovered that this had also left the rails, blocking the path of an up train from Derby to Leicester. As the latter engine was rerailed in a short time, the up train proceeded to Barrow, where it was divided, half going on to Leicester and the other half returning to Derby.[35]

By the end of 1841 the general state of the track had not improved, and in November that year Woodhouse asked to be relieved of his duties as Superintendent in order to devote himself entirely to engineering matters, mainly on account of the extremely bad state of the baulks between Leicester and Rugby.[36] This section apparently continued to give trouble until it was replaced by new track with cross sleepers at three-foot intervals carrying rails weighing 65 pounds a yard in 1844.[37] Bell was asked to take over the Superintendent's duties in addition to his position as Secretary, with a consequent increase in his salary. Woodhouse seems to have left the company soon afterwards to work on a survey of a railway between Leamington and Coventry; he then went abroad and died in 1855. In the wake of his resignation, on 1 March 1842 the management committee asked William Cubitt to inspect the line "at his earliest convenience" and to advise upon the best steps to be taken to obtain a replacement. No doubt as a direct consequence of this appeal engineering matters were henceforth placed in the hands of William Henry Barlow.[38]

Other accidents were caused by negligence on the part of the staff, there being cases of points wrongly set, of crossing gates not being opened in time for a train[39] (despite the clause in the original Act which stipulated that all gates were to be kept closed to the roadway except when required to pass road traffic), and at least one – a collision – due to the flagman with a ballast train having left his post.[40] Two other serious collisions clearly illustrate the absence of any properly safe traffic regulation. The first of these occurred in January 1841 between Spondon and Derby, when a goods train was run into by the mail train during fog at night, the crew of the latter being killed. The goods train had left Leicester with two engines an hour and a half in front of the mails, but a mile and a quarter short of Derby one of the engines failed with frozen feed-water pumps. While the guard went back six to seven hundred yards in order to protect the train, the sound engine towed the failed one onwards to Derby and, being unable to find a pilot engine and incapable of taking on the whole load unassisted, then had to return twice for the stranded vehicles, the mail train running into the last portion.[41]

The second serious collision occurred on Easter Monday 1842, when an up mineral train consisting of 16 wagons and a *break*, left Derby at 5.30 am and, after travelling at 12 mph, stopped for water at a tank two miles short of Leicester. After travelling a further half-mile the driver noticed that he was being overtaken but, although he accelerated from 5 to 6½ mph, the train was run into half a mile farther on. The following service was a luggage train with two engines. When the driver of the leading one saw the goods train ahead, he shouted back to the other driver that, as the goods engine had been taking water, there would be none left for them. The second driver acknowledged this message but, being otherwise engaged, failed to shut off steam with the leading engine after passing the water tank and, although the first engine was put into reverse, the driver was unable to prevent the inevitable collision, in which his fireman died. The Company already had a regulation that engines should not approach within 800 yards of another train travelling in the same direction on the same line,[42] but drivers had of course to rely upon seeing the potential obstruction in time.

RECEIPTS, MANAGEMENT, AND COMPETITION

The two half-yearly General Meetings of 1842 were characterised by lengthy discussions of running expenses. Figures given at the meeting on 23 February showed that these compared very favourably with those of the London & Birmingham Railway, including track maintenance by contract at £200 per mile, coke at 20s a ton, and 21 engine drivers paid 42s a week to operate the 19 daily trains.[43] At the next meeting on 13 August the gross receipts for the first six months of 1842 were quoted as £64,007 7s 3d, and the balance on the revenue account as £18,767 18s 2d, which enabled a dividend of only 1½ per cent to be declared. Numerous economies were thereupon suggested, but the discussions did not placate the worries of all of the shareholders, and on 2 November a sufficient number of them issued a notice calling for a Special General Meeting to be held on 18 November, its purpose being to appoint a Committee to examine the Company's finances. Although the Chairman announced that the Directors had already appointed a sub-committee, including John Ellis on behalf of the dissentients, to consider economies amounting to £4,500, a committee of shareholders was appointed by 4,024 votes to 3,309.[44]

Most of the economies which were immediately practicable were introduced by the Board before the Special Committee issued its report at the end of January 1843. The employees objected strongly to the reductions in both wages and the number of staff which were introduced, and went so far as to issue a public notice stating their case.[45] When it came, the report was very diffuse, and its several recommendations towards improving the financial management were presented to a Special Meeting of the shareholders in Derby on 14 February.[46] The motion "that the charges against the Directors had been satisfactorily answered" was carried by a small majority only.[47] Although the storm had been weathered, the Directorate did not come through unscathed, and at the next half-yearly meeting on 15 March their number was reduced to twelve. The reconstructed Board consisted of Messrs John Cartwright, T.E. Dicey, John Ellis, Douglas Fox, William Hannay, Lawrence Heyworth, William Evans Hutchinson, James Oakes, George Byng Paget, John Taylor, Samuel Waters and Henry Youle. A second dividend of 1½ per cent for the latter half of 1842 was then declared.[48]

Anticipating that it might be difficult to secure all of the through traffic between Derby and points south of Rugby from the Birmingham & Derby Junction Railway, whose route through Tamworth and Hampton had been in use for some ten months already, the Midland Counties Railway Directors approached their competitor for an agreed division of the traffic just before completion of the line between Leicester and Rugby.[49] No agreement was forthcoming and, despite a longer route, at 141 miles 48 chains against 130 miles 61 chains over the Midland Counties Railway, and the refusal of the London & Birmingham Railway to reduce its share of the receipts between Rugby and Hampton, the Birmingham & Derby Junction Company retaliated by reducing the through fares until its own share on first and second class journeys from Derby was only 1s 6d. Although it was anticipated that the two year old agreement with the North Midland Railway to book all through traffic over the Midland Counties route would limit the effect of the Birmingham & Derby Junction reductions, the terms covered a period of only seven years, terminable should *any other*

company offer lower rates and the same average journey time during three successive months. The North Midland Railway was already having difficulty in its relationship with the Birmingham & Derby Junction Company over the existence of this agreement and had been offered cancellation provided the latter Company raised its fares to the level of those on the Midland Counties route and made certain adjustments to the timetable.[50] The agreement appears to have been maintained, however, until it was outlawed by the Midland Counties Railway Act of 1840 at the instigation of the Birmingham & Derby Junction Company, upon a penalty of £50 per day.

Intensive competition for this through traffic between Derby and London caused a great deal of ill feeling and considerable financial losses to both parties. As there were no comparable reductions in *local* fares on the Birmingham & Derby Junction route, and counsel advised the Midland Counties Railway that this was *ultra vires*, application was made to the Court of Chancery for an injunction to restrain the Birmingham & Derby Junction Company from continuing with its through fares reductions. The application was heard before the Lord Chancellor on 8 August 1840, but was dismissed with costs under "the decided opinion that such proceedings were in accordance with the Company's Act and that it would be improper to prevent the public from enjoying the advantages of cheap travelling under the pretence that such interference was for the benefit of the public".[51] Having failed in all attempts to force the competitor to capitulate, the Midland Counties Board proposed a division of traffic at the end of August 1840, passengers and carriages to be sent at separate and stated times by each route. Obviously not anticipating full success, the Board also resolved that from 14 September the Company's Derby–London fares should not exceed those charged by the Birmingham & Derby Junction Railway and on that date the first class fare was accordingly reduced from 35s to 27s and the second class one from 24s to 18s. *The Derby*

Mercury remarked that the receipts for this single day almost equalled those of the previous week.[52]

In October 1840 the Midland Counties Railway nominated a Committee to negotiate with the Birmingham & Derby Junction Company as soon as the opportunity arose,[53] although nothing seems to have developed until September 1842, when the Birmingham & Derby Junction Railway proposed that the London & Birmingham Company should undertake the working of both routes to Derby. In view of their own parlous finances, the Midland Counties Board had already made overtures to the London & Birmingham Railway on these lines, for "running the carriages and to have the undertaking taken at valuation".[54] How different the future organisation of the railway system might have been if the London & Birmingham company had swallowed up these two lines and thus frustrated the progression towards amalgamation as the rival Midland Railway. Under the Birmingham & Derby Junction scheme the net receipts would have been divided, two thirds going to the Midland Counties Railway and one third to themselves, an arrangement which the Midland Counties Board decided it could not accept.[55] A counter proposal that, to avoid competition and provide the best possible public service, the Birmingham & Derby Junction Railway should relinquish all through traffic on agreed terms was made and another Committee appointed to negotiate.[56] Three months later, in January 1843, Henry Smith, the Birmingham & Derby Junction Chairman, suggested an immediate amalgamation, according to an assumed nominal market price of £40 for Birmingham & Derby Junction shares and £60 for those of the Midland Counties Railway. Although this idea was also rejected, negotiations were opened for amalgamation "upon fair and equitable" terms.[57]

From 13 February 1843 both Companies advertised identical fares between Derby and London, at 30s first, 20s second, and 14s third Class, with about six trains in each direction by either route.[58] From 3 April these were once again reduced on certain competing trains to 28s First and 19s Second Class and about the same time the London & Birmingham Railway agreed to book passengers from Birmingham to Derby through Rugby at 8s, 5s, and 3s.6d.[59] A special advertisement pointed out the advantages of using this route, with its direct connections into the Grand Junction and the Birmingham & Gloucester Railways at Curzon Street, the Birmingham & Derby Junction terminus at Lawley Street being at a lower level and less convenient.[60] Three months later the rates for through goods, cattle, and coke from Derby to Rugby were reduced in answer to cuts by the Birmingham & Derby Junction Railway,[61] and in July a further reduction in passenger fares of 2s was offered by one train each way between Derby and London.[62] In April it was once again decided to take legal opinion upon the practices of the Birmingham & Derby Junction Company in charging different rates for local passengers, and application was made in the Court of Queen's Bench for a *mandamus* to restrain the competition. Advice was received that the fares were both unfair and illegal, and a *mandamus* was issued which could be enforced in November 1843.[63]

In April 1843 negotiations with the London & Birmingham Railway were reopened, when the Midland Counties Railway asked its more powerful neighbour to consider taking over some of the coaches and the working of all through carriages over the line.[64] Even though nothing came of these particular negotiations there were certain occasions upon which London & Birmingham Railway engines penetrated the Midland Counties line, witness the permission granted in October 1843 for the London & Birmingham Company to convey, by its own power, 25 wagons of granite from Syston to Rugby.[65] This was of course permission for one particular journey and not for general running powers over the line.

The statement for the first six months of 1843, presented at the half-yearly General Meeting on 10 August, showed the gross receipts to be a little lower than in the previous half-year, at £62,324. With expenses of

£44,987, a net profit of £17,366 remained in spite of the fares war. After deducting £1,500 owing to the North Midland Railway, a dividend of £1 4s 3d per cent was declared.[66] In July wage reductions and other alterations resulting in an annual saving of £950 14s 8d were made,[67] and in September engine drivers's wages were reduced to 6s a day. A mixture of one part coal to two parts coke was tried in the goods engines to reduce fuel costs.[68] Despite the economy drive it was decided to carry out certain station improvements at the end of 1843. Lighting was improved and the platforms lengthened at Syston, Sileby, and Kegworth.[69]

Financial agreement had been reached on 26 December 1842 as to the use of the North Midland Railway station premises in Derby. Interest at 6 per cent on the £12,000 spent to provide the Midland Counties Railway accommodation was due to the North Midland Company, and there was to be some rearrangement of freehold between the two Companies as well as a rental payment by the North Midland Railway towards the goods office built on Midland Counties land, which would be removed if and when required.[70] In the spring of 1843, however, the North Midland Company furnished a new abstract of the accounts due to it and an agreement for the use of the station was finally settled on 20 June,[71] the Midland Counties Company promising to pay the balance of £3,404 5s outstanding, free of interest. In future the Company would pay £3,590 annually, made up as set out in Table 9. A sum of £4,293 12s 6d was paid in July 1843 in settlement of the working account to 30 June 1843.[72]

FURTHER ENGINEERING WORKS AND ACCOMMODATION

During 1843 the Company decided to sell to the London & Birmingham Railway all of its land on the south-east side of the two lines at Rugby at prime cost if the latter would build and let to the Midland Counties Railway six cottages at a rent equal to that charged to its own servants.[73] In July an indictment was served upon the Secretary from the Parish of Rugby complaining that, in conjunction with the London & Birmingham Railway, the Company had stopped up an ancient footpath.[74] It was withdrawn in April 1844 upon condition that the Midland Counties Company built an archway for foot passengers underneath its line and a new path beside the railway. Mr Broughton Leigh would only agree to the construction of this new footpath if he and his family were allowed to enter the station from the road under the railway at the end of the path, so avoiding the inconvenience of going round through the Town.[75]

In order to facilitate the building of the bridge across the river Trent, the original Act had allowed replacement of the existing weir by a new one of ashlar stone farther east, downstream, at the Railway Company's expense. As the bridge was erected without destroying the old weir the Company naturally tried to avoid the now unnecessary cost of building the new one afterwards. The Trent Navigation Company, as owners of the old weir, at once determined to force the Railway Company's hand, however, and the Secretary and two of the Directors were served with a *rule nisi* from the Queen's Bench, requiring them to show why the work had not been done as required. The Company countered by suggesting the cutting of a new canal direct from the mouth of the Erewash Canal into the river Soar, but this was resisted by the Navigation, as was any delay on

TABLE 9   Payments for accommodation at Derby

Interest, at 6% on £12,000	720
Managing goods, etc.	300
Toll over bridge and all repairs thereof	450
Pumping water	120
Every other expense	2,000
	£3,590

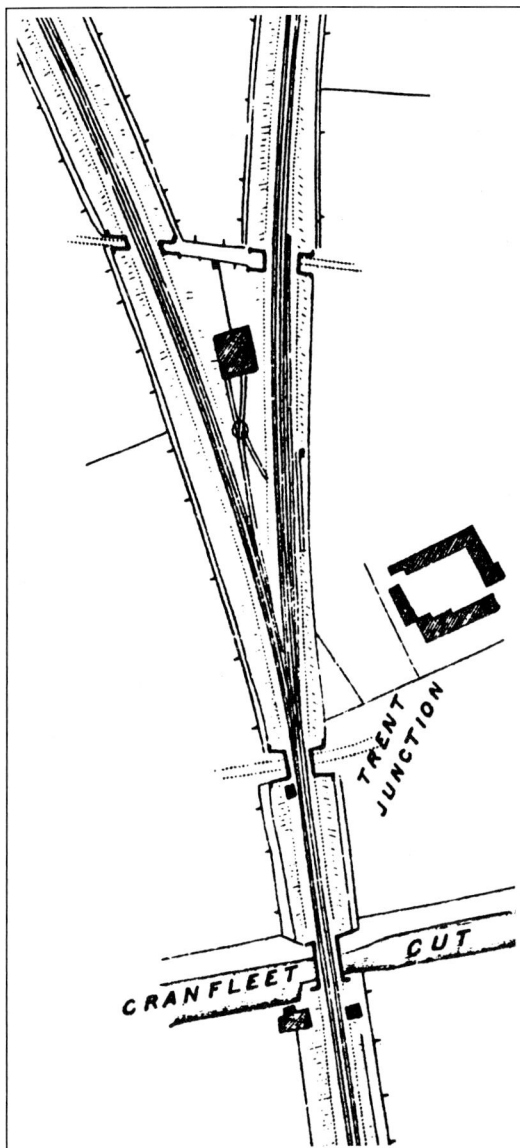

beginning the replacement weir until the spring of 1841. Further legal action was taken by applying to the Court of Queen's Bench for a *mandamus* to enforce the building of the new weir.[76] Although the provisions of section 105 of the original Act were repealed by another Act of 1842,[77] the Company was still required to construct a new weir, not more than 200 yards below the eastern side of the bridge. Although Cubitt was asked to furnish plans and specificiations for the work in February 1842,[78] no progress seems to have been made until the Trent Navigation Company's solicitors complained of continued inaction in March 1843, upon which Cubitt was asked to proceed without further delay.[79] One month later he recommended the appointment of a Mr Carr to assist Barlow on the weir so that the latter could concentrate upon the railway, but by this time contracts for the necessary timber framing, at £125 including removal when finished with, had been arranged. Mountsorrel Granite, Millstone Grit, and various timbers were used in the construction itself, and by the end of August Barlow was told to arrange for the removal of the old weir.[80] In September the works were reported to be "nearly out of the Contractor's hands".[81]

Further engineering work was required towards the end of 1843 to remedy the flooding caused by the embankment at Loughborough, some of the culverts being lowered to improve the drainage. Early in 1844 the driver of an up mail train received 20s for reporting a slip in Leire cutting, which the Contractor was able to remove without closing more than one line.[82] Another slip of 20,000 cubic yards at Gill's Corner, near Rugby, was removed later in the same month without causing any delay to traffic.[83]

One point on which little information is available concerns the arrangements for combining and dividing trains at *Trent Junction*. From the date on which the line was opened as far as Leicester, most of the trains served both Nottingham and Derby and were divided or joined together at the southern angle of the triangle, though there were no proper

32. Site plan of platform at Trent Junction
The plans for the Midland Railway Bill of 1848 indicated a very small platform alongside the southbound ''up'' line from Nottingham just north of the junction at Trent, together with what appears to have been a small locomotive provisioning establishment served from the southbound line from Derby. Nottinghamshire Records Office.

station facilities at this point. The importance of Nottingham declined somewhat after the inauguration of through traffic between London and the North through Derby, the withdrawal of the Nottingham section of the mail service having already been noted. In June 1841 a newspaper correspondent was complaining of the lack of through carriages for Nottingham on some of the London trains, the writer having been asked "by a smart dapper gentleman in a lemon coloured waistcoat" to change carriages at Leicester. He went on to say that "at the point of separation of the Derby and Nottingham lines, the third class passengers were all turned out into another carriage ... and kept waiting. Had the weather been bad, they would have been drenched to the skin by standing in the open air, and on wet ground".[84]

It is evident that any accommodation provided for the passengers at Trent Junction must have been very meagre, especially when the superior classes were asked to change at Leicester. The lack of shelter is patently evident, and there does not even seem to have been much of a platform, although the coaches were generally provided with sufficient steps at that time to make climbing down to ground level perfectly feasible, if somewhat acrobatic. It comes as no surprise therefore to learn that in April 1842 a man alighting from the 7.30 pm train from Nottingham was run down by the engine as it was running round the train prior to pushing it onto the rear of the Derby portion before the combined loads continued onwards to Leicester.[85] Why he needed to cross the path of the engine is not clear but this arrangement seems to have been allowed to continue. In June 1843 it was decided to issue special tickets at 5s first class and 3s 6d second class between Nottingham and Derby "via the Junction".[86]

Even at this date, four years after the opening of this section, competition from road services was still keen, an agreement with Messrs Hardy & Son to collect and deliver parcels in Nottingham for £2 10s a week

expressly forbidding any connection with competing road services.[87] Two months later a new service of four coaches each way every day, at 2s 6d inside and 1s 6d outside, was advertised as "saving the Unpleasantness, Expense, and a Half Hour's delay in going down to the station, besides the danger of losing your luggage and Travelling by Steam, all avoided".[88]

GOODS AND MINERAL SERVICES

Little information is available on the number of goods trains and the methods of operating these services. On 5 May 1840 George Stephenson wrote to the Board enquiring their terms for carrying coal and lime, and the Locomotive Engineer and General Superintendent were ordered to work out the costs for conveying minerals of all sorts and heavy goods at a nominal speed of seven to ten mph.[89] During the following summer the railway began to make its mark upon the local carriers, two of whom announced that they were sending traffic by rail instead of in their own road wagons, since it was already "quicker by rail"! Some of their horses were being sold off. In October 1841 *The Leicester Chronicle* observed that a considerable traffic in farm and market garden produce had developed between Leicestershire and both London and Yorkshire.[90] The existence of an appreciable through goods traffic to the North may be inferred from the presence of the Company's representative at a meeting between the Manchester & Leeds, the York & North Midland, the North Midland, the Hull & Selby, the Great North of England, and the Birmingham & Derby Junction Railways, which was held at Normanton on 23 June 1843. The meeting decided, amongst other things, to appoint at joint expense someone based at Normanton to superintend the use of wagons, couplings, and sheets by the several different companies involved.[91]

At the half-yearly meeting in April 1841 it was announced that preparations were still being made for the transport of coal and minerals, presumably southwards, as a coal wharf had been opened in Leicester during March.[92] Despite its earlier origins amongst the coalmasters of the Erewash Valley, the Midland Counties Railway did not serve any coalmines directly, and any coals carried must have been received either from the North Midland Railway or, less likely perhaps, by transhipment from road or canal. Certainly some of the Midland Counties traffic was transhipped *onto* the canals, the prospects of increasing traffic leading to an expansion of the accommodation at Rugby Wharf towards the end of 1843, so that four boats on the Oxford Canal could be loaded at once.[93] It was also decided to charge 1s per ton for coal in owners' wagons from Derby to Long Eaton, the owners having to arrange with and pay the Company's servants for running their wagons down the incline and loading the boats in the basin off the Erewash Canal.[94] It is not clear as to where this traffic would have been consigned, since farther distances both east and south were available by rail before transhipment became absolutely necessary, and it would hardly have been sent north towards the Erewash Valley collieries. In July 1843 a reduced rate of ¾d per ton per mile was offered to anyone sending more than 8,000 tons annually from Derby to Rugby, presumably to encourage coal traffic from the North Midland Railway southwards along the Oxford Canal.[95]

A little earlier it had been decided to reduce the rate for cattle wagons between Derby and Rugby from 30s to 26s,[96] and the minimum wagon load from Leicester to Nottingham from two and a half to only two tons.[97] Goods train loads were not heavy for, by way of example, it was decided in February 1844 that trains between Derby and Rugby should be limited to 20 loaded wagons plus the *break* during the winter months.[98] Not all of the wagons were yet fitted with brakes, it being ordered that this device should be added to all of the double box coal wagons in March 1844,[99] though whether these carried removeable container bodies to facilitate transhipment onto the canal is not known. A Mr Machon was, however, requested to return

the bodies for coal wagons lying at Pinxton Wharf by canal to Long Eaton in February 1842.[100]

During the winter of 1841 Derby Town Council invited the Royal Agricultural Society to hold its 1843 Show in the town,[101] a suggestion which obviously met with approval, since the event was duly held there in July. During the period that it was open, the Midland Counties Railway advertised special rates for wagons loaded to Derby,[102] and the London & Birmingham Railway agreed to provide through bookings by additional trains both from London and from Birmingham. In its issue of 15 July, *The Illustrated London News* gave the event extensive coverage. "Derby Railway Station", it stated, " – first by universal consent in the empire, or indeed in the world – claims special attention. Its prodigious extent, its incomparable plain form, its light but beautiful roof, its refreshment-room, its fine hotel, and the admirable manner in which its immense transactions are conducted, must fill every stranger with surprise and admiration". Interesting views were given both of its exterior and interior.

EXCURSION TRAFFIC

From its inception, the Company seems to have been anxious to take advantage of any opportunities for attracting excursionists by offering reduced fares and by providing extra trains, either by private arrangement or for the public at large at the Company's own initiative. Some of these arrangements have already been mentioned, and it is commonly accepted that Thomas Cook's first venture of this kind comprised a special train between Leicester and Loughborough on 5 July 1841 for *the friends of temperance* to attend a Quarterly Delegate Meeting and Gala at a special third class return fare of only 1s. The 570 participants were packed into nine open third class carriages and in Loughborough a band led the procession into a Park, where Cook had provided refreshments. After another procession, and games and dancing in the even-

33. A ticket for a rail and water excursion from Rugby to Matlock issued by the company on Wednesday, 22 June 1842. The heraldic device adopted by the Midland Counties Railway Company, comprises the coats of arms for the four counties through which its lines passed, i.e. Derbyshire, Nottinghamshire, Leicestershire, and Warwickshire. Original courtesy of Glyn Waite.

ing, the party returned to be met by great crowds in Leicester.[103] During the summer of 1842 the Company's Secretary organised several excursions to Matlock, at first from Leicester, the passengers being conveyed by train to Ambergate and thence by fly-boat along the canal to Cromford. About 700 passengers were carried in a 23-coach train, and finished their journey from Cromford on foot.[104] A week later, first and second class passengers on a similar outing from Nottingham and Derby were provided with an omnibus for the last stage of the journey, at inclusive fares of 8s first and 5s second class from Nottingham.[105] At least one similar excursion was run in the following year, with a slight increase in the fares.[106]

Early on Wednesday morning, 24 August 1842, a special train took passengers from Derby to London at a fare of £1 15s first and £1 5s second class, returning two days later.[107] It served most of the principal stations. Although contemporary reports indicate that in most cases the public response was overwhelming, the chief exception being one trip from Birmingham to Matlock in 1842, the Company appears to have had a change of heart, and no further examples have been found. Cheap fares by normal services were first introduced on specified trains between any two stations on the Company's routes at Whitsuntide in 1843,[108] and day return tickets were introduced generally from 1 January 1844 between Derby and the principal stations to Leicester at about one third reduction from the normal fares.[109]

5

Extensions and Amalgamation

During the operating life of the Midland Counties Railway Company several connecting schemes of varying importance were to appear – including, in 1842, a line between Northampton and Leicester which was to be worked by horses. There was a spate of railway schemes during 1844, which included lines from Kings Lynn to Leicester and from Banbury to Rugby, but of most importance were the Derby–Manchester and Nottingham–Lincoln projects.

DERBY TO MANCHESTER

There was considerable activity amongst promoters for direct lines between Derby and Manchester during the winter of 1840–41, and reports appeared under a number of different titles (though possibly describing the same project). The London & Manchester Railway, the London & Manchester Direct Railway, and the Manchester & Derby (Churnet Valley) Railway certainly all seem to have intended to follow the Churnet Valley and make use of the Midland Counties route between Derby and Leicester, the last having the support of this Company. The scheme first mentioned in October 1840 was for a line from the Northern & Eastern Railway at Broxbourne, through Bedford to Leicester, with a branch to Tamworth. The main line was to continue from Derby through the Churnet Valley to Manchester.[1]

In December 1840 the London & Manchester Railway – designed by Remington – was described as following a course at its southern end very similar to that later taken by the Midland Railway Company's own London Extension.[2] In the following month a "newly projected" London & Manchester Railway was described as following a very similar route, but with Rastrick as its Engineer.[3] At a meeting in favour of the London & Manchester Direct scheme in Cheadle during February 1841 a Mr Cattlow, supposedly speaking on behalf of the Midland Counties Railway, stated that the Churnet scheme was not yet dead,[4] and later in the month Rastrick reported that he found upon examination that Remington's route was quite practicable.[5] Until 1824 notice of application and deposition of plans for second class local Bills, into which class of legislation canal and railway plans fell, were required during the late summer months of August and September, such deadline being then extended into the subsequent autumn months of October and November. While this still applied to all other Bills in this category, in 1836 Parliament, against the advice of every engineer of repute, decreed that in future railway plans must also be deposited during the previous February or March. Whatever the reason, and several were advanced, the results, prepared under the worst possible weather conditions, were chronically imperfect and hastily compiled; it was often necessary to introduce amendments into the autumn plan where there had been any significant land developments in the interim. Until 1842, when this order was rescinded and deposition of plans reverted solely to the end of November, an additional burden had been placed in the way of railway promoters. Parliamentary notice thus appeared in the spring of 1841 for the Manchester & Derby (Churnet Valley) Railway under the auspices of the Midland Counties Company's new solicitors, Berridge & Macaulay of Leicester.[6]

After several optimistic meetings, no more was heard of any of these schemes until the end of 1843, when the Churnet Valley project was revived and the Midland Counties Railway paid Rastrick the £25 owing for his plans.[7] After a deputation had attended a meeting in May 1844 a prospectus was issued, naming a provisional committee of 23 persons including several of the Midland Counties Directors and suggesting a capital of £1,000,000. The route proposed by the Manchester & Derby company's new Engin-

eer, George Watson Buck, ran from the terminus of the Macclesfield branch of the Manchester & Birmingham Railway through Leek and Uttoxeter to join the Birmingham & Derby Junction Railway two miles south of Derby, a total route length of 46 miles.[8] Later amendments were suggested, to connect with the North Midland Railway at Duffield, north of Derby, and for branches to both Hanley and to Tamworth,[9] and another Parliamentary Notice was issued in November 1844. Although this line was never authorised or built, it had some influence upon the route selected by the North Staffordshire Railway, which was under consideration at the same time[10] and eventually obtained its Acts in 1846. The more direct routes for Midland Railway traffic between Derby and Manchester were not to be built for some years hence.

34. Initial plans for the Lincoln extension offered alternative routes through the Nottingham Meadows. This one, of 1840, shows the line taking off directly from the terminus of the Midland Counties Railway. Nottinghamshire Records Office.

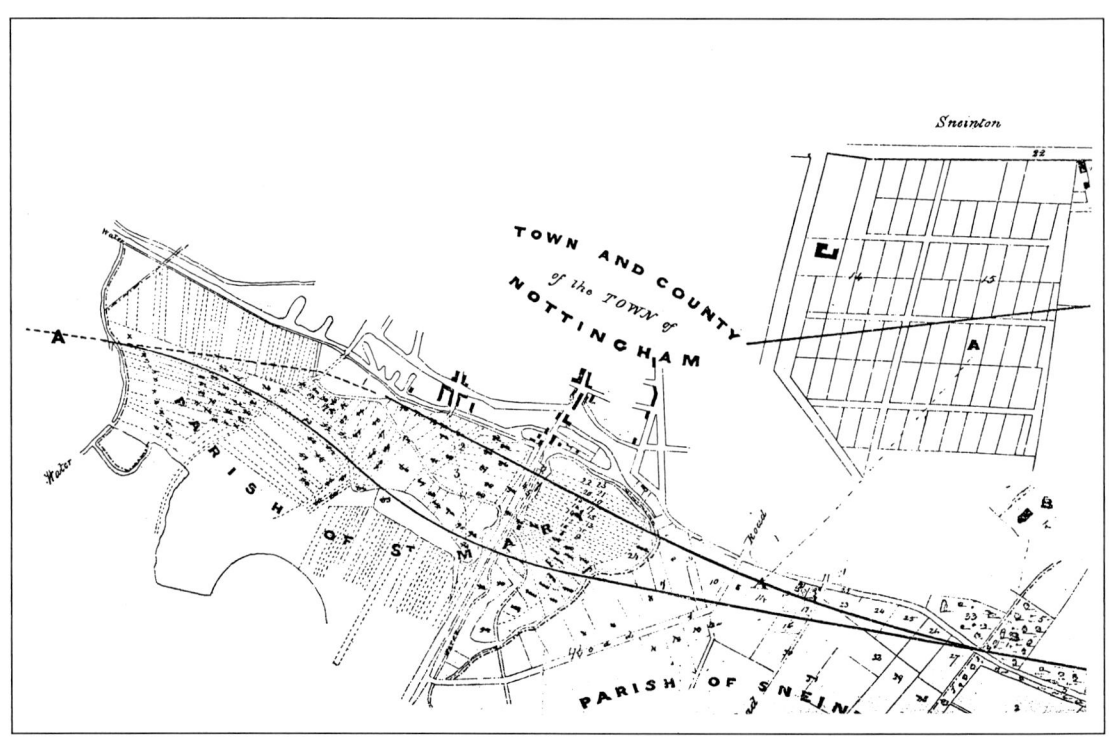

NOTTINGHAM TO LINCOLN

The construction of a line between Nottingham and Lincoln was first proposed in 1840 and the scheme was revived again in 1842, but it was not until 1844 that the Midland Counties Railway seriously considered such an extension and the preparatory work began in earnest. In February of that year, the Board ordered that Messrs Barlow and Dabbs should make a survey, and decided to ask Robert Stephenson to act as Consultant.[11] Applications for shares, at £25 each, were invited from both the existing shareholders and from the general public.[12] Stephenson replied that he had for some years declined to act as a Consulting Engineer, regarding such a position as "merely nominal, involving responsibility without either credit or profit".

The Company thereupon decided to appoint him sole Engineer, with full responsibility for the project[13] and, upon receipt of his estimate of 1 March for construction at £10,500 per mile,[14] a prospectus was issued.[15]

A capital of £350,000, divided into £25 shares upon which a deposit of £2 10s each was required, was proposed for the nominally independent Nottingham, Newark & Lincoln Railway. The Provisional Committee of 13 members was drawn largely from Midland Counties Railway Directors, together with a few local men. A line nearly 33 miles long was to begin alongside the Midland Counties terminus in Nottingham and pass two miles from Southwell to reach Newark and Lincoln, with a connection into the projected Lincoln & Wakefield line. Building costs over the very level countryside were expected to be low and

35. Parliamentary plans for the final conception of the Lincoln extension, lodged in November 1844, were the first to indicate, albeit sparsely, the layout of the Midland Counties station in Nottingham. Queen Street had been completed for Victoria's visit during the previous year and the West Croft canal already laid out by the Corporation. Nottinghamshire Records Office.

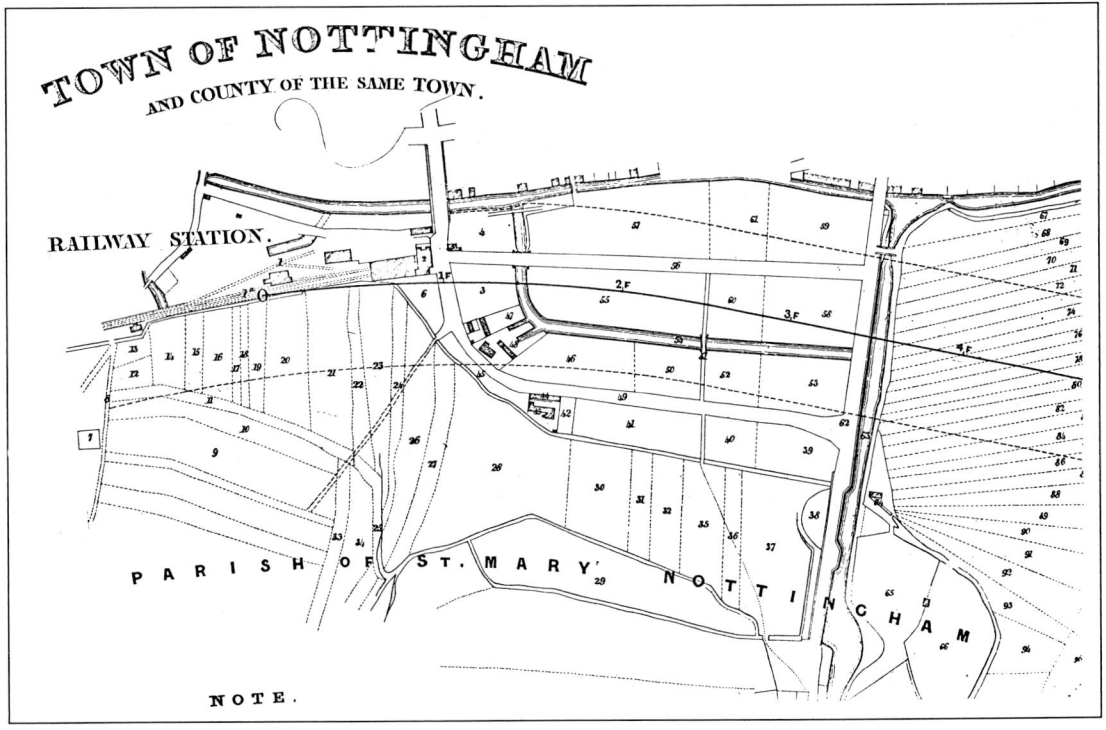

dividends good, anticipated working expenses being only 40 per cent of a gross revenue expected to top £40,000. Completion was expected (and in fact achieved) by August 1846, and it was proposed that the line should be leased to, or worked by, the Midland Counties Railway, this being a more economical arrangement than the provision of separate rolling stock.[16] In March it was announced that applications had been received for double the number of shares to be allotted.[17]

The first meeting of the provisional committee of the Lincoln Company was held in Nottingham on 9 April 1844, and in May the Midland Counties Railway guaranteed all expenses already incurred by the promoters, recommending the shareholders in the amalgamating companies to make the line an integral part of the Midland Railway upon the terms already proposed to the Lincoln & Wakefield group.[18] George Hudson duly announced at the first General Meeting of the new Midland Railway Company in July that applications to Parliament were proposed to make both of these lines, from Lincoln to Swinton and to Nottingham, together with another line, between Lincoln and March, designed to fend off the competition from other East Coast projects.

OTHER PROJECTS

Back in 1841, at a Special General Meeting of the Midland Counties Railway shareholders on 5 November, it was decided to subscribe £3,000 each year for a term of ten years towards the construction of the Newcastle & Darlington Junction Railway, in conjunction with the other companies interested in completing railway communication between Newcastle and the South. Sums were also guaranteed by the North Midland Railway, the York & North Midland Railway, the Manchester & Leeds Railway, and other companies operating north of York.[19] In December 1843 the Board agreed to relinquish this agreement if the Newcastle & Darlington Junction Company so requested.[20] About the

same time, the Joint Management Committee's suggestion that a lease of the Leeds & Bradford Railway might be obtained was not taken up.[21] Also during 1843, the Erewash Valley Railway, now being promoted independently, was given permission to make a junction with the Midland Counties Railway at Long Eaton.[22] Nothing came of the Direct London, Manchester & York Railway's branch from Bedford to Leicester, promoted in 1844.[23]

AMALGAMATION AND ITS EFFECTS

By the middle of 1843 it was already realised that the ruinous competition which had raged between the Midland Counties Railway and the Birmingham & Derby Junction Railway could not continue indefinitely. Although there is little doubt that the Birmingham & Derby Junction Company was the heavier sufferer, the Midland Counties Company was now prepared to make greater efforts to reach a satisfactory settlement. Early in July the Chairman, together with three of the Directors (Ellis, Taylor and Youle) met the Chairman of the Birmingham & Derby Junction Railway in Birmingham, when the latter party proposed a temporary cessation of the rate war, along with an agreement that passengers should be booked exclusively via Hampton by two trains in each direction and via Leicester at other times, pending application for an Act to unite the two companies. The Midland Counties party would not agree to defer the actual terms of union until after the Act had been obtained,[24] but decided to press on with the negotiations.

On 1 August, however, Newton, Hudson, and Waddingham of the North Midland Company brought forward a proposal for the union of both companies with their own,[25] to which the immediate reaction was that the outstanding differences with the Birmingham & Derby Junction Railway should first be settled.[26] The latter party agreed to the North Midland proposal and consequently refused to entertain any separate agreement with the Midland Counties Railway.[27] Matters were then taken out of the Directors' hands by the circulation to all shareholders of a letter from Hudson, Waddingham, Holdsworth and Wilson, favouring the projected union, to which the Board retaliated by circulating an answer.[28] On the morning of the half-yearly meeting on 10 August 1843 Hudson, Waddingham, and Holdsworth met the Board to announce their intention of moving a resolution in favour of amalgamation upon the terms proposed by the North Midland Railway;[29] they had presumably purchased sufficient shares in the Midland Counties concern to allow them to do this. When the resolution was introduced at the meeting a lengthy and noisy discussion ensued. After Hudson had suggested that amalgamation would increase the joint profits to upwards of £35,000 per annum, largely by reducing expenses, the meeting finally decided, against the strenuous opposition of the Directors, to appoint a Committee of certain Directors and shareholders to confer with the other two companies and to report back to a special meeting.[30]

Since this committee insisted that hostilities against the Birmingham & Derby Junction Railway should immediately cease, the Board at once resolved to approach the North Midland Railway Committee for its consent, and to draw up the terms upon which an acceptable truce could be arranged.[31] When these Committees met on 21 August it was recommended that all Midland Counties and Birmingham & Derby Junction charges should revert to those current before the onset of the latest outbreak of fares reductions on 13 February, which terms were approved at the next Board Meeting, only the chairman being opposed.[32] The Committees of all three companies then met on 30 August, when it was decided to draft out an agreement for the joint working of the three undertakings, with subdivision of the receipts until an Act of Parliament could be obtained to ratify the proposed amalgamation. In order that this might be obtained during the next Session, the agreements were submitted to Special General Meetings of the shareholders of each

company after prior circulation of the terms.[33]

The Midland Counties Railway meeting was held on 21 September 1843, after the proposed terms had already been agreed by the shareholders of the other two companies. They comprised consolidation of all properties into a single company as soon as possible, the transfer of all liabilities up to a total of £1,855,000, which included the Midland Counties Railway share of a £581,000 debt, limitation upon further debts by any party, the assumption of liability for all agreements previously entered into, and various financial and administrative matters. Of a total share capital of £5,158,000, an amount of £1,275,000 was to represent that of the Midland Counties Railway, each shareholder to be allotted new stock equivalent to his previous holding. Holders of the £20 shares were to receive a guaranteed dividend of six per cent and the remainder equal proportions, except on Birmingham & Derby Junction Railway £100 stock, which would carry 27s 6d less annually. The new Board was not to exceed 15 directors, and in the meantime a committee of three directors from each of the existing companies would be nominated to bring the amalgamation into effect by promoting the Act as soon as possible. Subject to the sanction of the existing Boards, this Committee would henceforth manage the three enterprises as a single system.

These proposals were carried by a majority of 115 to only 11 votes amidst considerable uproar, during which Mr Dicey, in the chair, demanded a poll on behalf of absent voters, he being armed with the votes of such dissenters. After he had been removed from the Chair, a count still found in favour of the union, but Dicey threatened to oppose the Bill in Parliament.[34] Taylor, Ellis, and Hutchinson (two of whom were later to become in turn chairman of the new Midland Railway Company) were appointed to represent the Midland Counties Railway interest on the caretaker committee,[35] and economies were soon introduced. On 3 October it was recommended, amongst other things, that only one pilot engine should henceforth be pro-

36. The Queen at Nottingham Station, exterior
On arrival from Chatsworth by rail over the North Midland and Midland Counties Railways, the Queen left Nottingham Station for Belvoir by coach, passing along the newly opened and highly decorated Queens Road. *Illustrated London News,* courtesy Nottinghamshire County Libraries.

vided at Derby, and that the London & Birmingham Railway should be asked to provide the only pilot engine at Rugby. The three enterprises should be considered as a single unit by the Railway Clearing House, and the rolling stock of all three systems would now be used as most convenient. In order to enable the use of North Midland and Birmingham & Derby Junction Railway engines, some of the guard rails had to be removed from certain Midland Counties bridges.[36] It was also proposed that through carriages should be introduced between Nottingham and Birmingham, and that the Midland Counties Railway could work the trains between Nottingham and Derby with the same engines as the Rugby trains, presumably by adding and detaching carriages and wagons at Trent Junction.

As has already been noted, from 24 September the fares on the two routes between Derby and Rugby were restored to the levels current until the previous February.[37] During November a reduced rate

for coals travelling 50 miles or more over the combined systems was introduced,[38] and the Joint Management Committee recommended the provision of five new first class carriages and the introduction of a new first class train between London and Newcastle on the opening of the line beyond Darlington.[39] Several train alterations made during 1843 included the withdrawal of one of the services to and from Rugby as a result of the end of the competition with the Birmingham & Derby Junction Railway.[40]

By modern standards the trains were still very slow, first and second class services taking six to seven hours between Derby and London, some involving a change at Rugby. The lower orders were compelled to spend as much as 8¾ hours on the up journey and no less than 10½ hours in the down direction. In reply to one of the shareholders at the first General Meeting of the consolidated Midland Railway Company, held on 16 July 1844, who alleged that third class passengers were delayed for three hours at Rugby and again at

37. For the Queen's visit on 4 December 1843 the arrivals side of Nottingham Station seems to have been totally given over to a viewing stand. The shallow depth of the platforms is apparent in this view. *Illustrated London News,* courtesy Nottinghamshire County Libraries.

Roade in order to discourage them, Hudson replied that the London & Birmingham Railway already carried them at a rate in excess of the 12 mph required by Parliament.[41] Despite such practices the third class receipts, at 1d per mile, amounted to £16,693 during the year ending 30 June 1843.[42]

Two bye-laws of some interest were approved in October 1843. One dictated that "passengers at road stations will only be booked conditionally, i.e. if there is room; if there is not room for all, those going furthest will get priority". The second established a penalty of up to 40s for smoking, with the possibility of eviction from the Company's premises and forfeiture of the fare, the same penalty applying to drunkenness and obstruction of the company's officers.[43] During the latter half of 1843 and in 1844 there was a considerable correspondence with the Postmaster General concerning carriage of mails both by special train and under the charge of the guard on normal public services, but there is no indication of the outcome.[44] In March

1844 the threat of a general strike by the miners prompted the Board to lay in a stock of 500 tons of coke and 1,000 tons of coal.[45] As has already been amply demonstrated, the interests of the coal trade and railways remained close from the very inception of the latter.

An event of some importance took place on 4 December 1843, when Queen Victoria arrived at the Nottingham Station with Prince Albert and their full suite of retainers on a journey in part by rail from Chatsworth to Belvoir. They were received in style by the Earl of Scarborough as Lord Lieutenant, the Mayor, and the Commanding Officer of Her Majesty's Northern Forces, and conducted to an apartment suitably fitted up for such a great occasion. The party then left by carriage beneath the triumphal arches which had been erected over the newly made Queen's Road, now connecting Carrington Street with London Road, on its way to Belvoir Castle. After a short break, during which the royal couple joined a shoot and a well patronised meeting of the local hunt, the party returned by way of Melton Mowbray to take the train from Leicester to Watford, where their horses had been stabled ready for the last lap of the journey. On arrival in Windsor the Queen appeared to have experienced little fatigue from the lengthy journey of nearly 120 miles from Belvoir, the rate of travelling by railway having averaged twenty-five miles an hour with remarkable precision throughout the route. The same carriages formed the train during the whole of the visit and, as was the usual custom whenever her Majesty travelled by rail, a pilot engine preceded the train at a distance of a mile to ensure full security.[46]

After the formal notice of intention to proceed with the application to Parliament had been issued on behalf of all three companies,[47] it was agreed that the new Directorate would be made up of six North Midland Railway members, five from the Midland Counties Railway and four from the Birmingham & Derby Junction Company.[48] Following a ballot taken in January, Ellis, Hutchinson, Taylor, Waters, and Youle were chosen from the Midland Counties Board.[49] In February the Joint Managing Committee chose Grisborn and Dougdale to be the MPs whose names were to be placed on the petition for the Bill.[50] Opposition to the amalgamation now developed amongst some of the Midland Counties proprietors, and on 2 February 1844 a meeting of the holders of the one-fifth shares, on which only £2 had been paid up, was held in Leeds. It was their contention that the terms of reference were unjust, and they proposed to oppose the amalgamation unless some provision was made to cover the payment of further calls on their shares. Subscriptions were agreed towards the costs involved and in due course a petition was presented to the Parliamentary Committee considering the Bill.[51]

The Midland Railway Consolidation Bill went into the committee stage at the end of February, and the Commons committee reported on 26 March. Six petitions had been heard from persons who considered that the Bill was not in their interest, including one from certain coalowners in the Erewash Valley complaining of the proposed method of levying tolls.[52] The Committee made several amendments, which were agreed to, and the Bill was engrossed on 29 March. It was read for the first time in the Lords on 1 April and passed with all amendments on its third reading on 6 May. Meantime the Midland Counties Railway shareholders met to consider the draft of the Bill on 16 April, when it was unanimously approved, with a minor amendment on the method of choosing a Chairman at General Meetings. Similar amendments at meetings of the North Midland and Birmingham & Derby Junction Railway proprietors were, however, defeated, the Bill being approved as it stood. It received Royal Assent on 10 May 1844, from when the Midland Counties Railway ceased to exist upon its vesting in the new Company. A final dividend of 2⅛ per cent was paid on the results of the last six months of independent working.

That railways were still a source of wonder is evidenced by a report in the same month of the large crowds which assembled at Derby Station to watch the departure of "a most extraordinary luggage train" drawn by three engines.[53] From 1 July 1844 fully unified working of the new Midland Railway system began. The first of the great railway amalgamations had become effective (to be followed two years later by the creation of the London & North Western Railway), and the Midland Railway began an eighty-year career that would take it into all four countries of the United Kingdom and make it one of the three greatest of the British railway companies.

6

Locomotives [1]

When the Company was first promoted in 1832 there was little experience of locomotive haulage over long distances. The engines used on the Stockton & Darlington Railway from 1825 were somewhat cumbersome and those on the Liverpool & Manchester Railway in 1830 rather lightweight. The proprietors went therefore to the Engineer of the Glasgow & Garnkirk Railway, which had also been using locomotives for some fifteen months. He reported that these engines had been constructed similarly to those in use on the Liverpool & Manchester Railway, having cost each, complete with tender, about £650. The weight of the complete unit was around 6 tons, exclusive of water, the power of each one amounting to some eighteen horses. The quantity of coke consumed in moving one ton over one mile had proved to be 1.164 lb, the cost of which, at 15s 6d per ton of 2,240 lb, amounted to .097d. The expense of attendants, oil, hemp, etc. amounted in addition to .18d, making a total of .277d, or a little more than one farthing per ton per mile. The engines travelled at the rate of from eight to ten miles an hour, drawing from twenty to twenty-four loaded waggons weighing four tons each, or between eighty and ninety-six tons gross weight. The rate *charged* on this railway for the moving power for each net ton was .375d, or a little more than one third of a penny per ton per mile. A summary of this report was published with those of Joseph Glynn and William Jessop when the prospectus was issued in November 1832. Glynn's report is set out in full at Appendix 3.

A LETTER FROM BUTTERLEY IRONWORKS TO THE DIRECTORS

13 March 1838

"Agreeably to your request, we submit to you the following proposal for supplying the loco-motive engines, which will be required for the railway, and we need not point out the advantage and convenience that will be experienced in having an establishment for their supply and repair in the immediate neighbourhood of the railway, where personal superintendance might be given and immediate access might be had to the manufactory to replace injuries by accident or otherwise, and where suitable engineers might be instructed and provided, so as to ensure the Company efficient services.

The machinery we would undertake should be made with the latest improvements, and of workmanship, material and construction equal to those supplied to the London and Birmingham Railway Company, and at the prices of the manufacturers of good reputation for corresponding work and material. We have already provided existing establishments for manufacturing the loco-motive engines and are practically acquainted with them. As it has been suggested that it would be desirable to contract for the maintenance and working of the engines on the plan adopted by the London and Birmingham Railway Company, we are willing to entertain the subject and to enter into a treaty on that

NO	NAME	NUMBER OF WHEELS & DIAMETER	CYL DIA	VALUE	
				ENGINE	TENDER
1	Bee	2 × 5′0″ & 2 × 3′0″	11″	700	80
2	Hercules	2 × 5′6″ & 2 × 3′6″	13″	850	90
3	Hawk	5′6″ & 4′0″	12″	1,050	130
4	Sunbeam	2 × 5′6″ & 2 × 4′0″	12″	1,000	135
5	Wizard	2 × 5′6″ & 2 × 4′0″	12″	700	55
6	Hecate		12″	700	120
7	Lion		12″	1,000	120
8	Tiger	2 × 5′6″ & 2 × 4′0″	12″	1,050	150
9	Vulture		12″	1,050	135
10	Eagle		12″	1,050	150
11	Leopard	2 × 5′6″ & 2 × 4′0″	12″	1,020	130
12	Panther	2 × 5′6″ & 2 × 4′0″	12″	800	110
13	Reindeer	2 × 5′6″ & 2 × 4′0″	12″	1,050	150
14	Antelope	2 × 5′6″ & 2 × 4′0″	12″	850	125
15	Unicorn		12″	850	125
16	Cerberus		13″	1,275	160
17	Caliban	5′6″ & 4′0″	13″	1,275	160
18	Basilisk		13″	1,150	160
19	Phantom		13″	1,275	160
20	Lightning	5′6″ & 4′0″	12″	1,000	135
21	Lucifer		12″	700	50
22	Hurricane		12″	1,050	135
23	Firebrand		12″	1,050	150
24	Rainbow		12″	700	120
25	Sirocco		12″	1,050	150
26	Dragon		12″	700	120
27	Scorpion		12″	700	100
28	Hornet		12″	700	120
29	Wivern		12″	800	130
30	Vampire		12″	850	135
31	Lynx		12″	850	135
32	Centaur		13″	1,275	160
33	Hydra		13″	1,275	160
34	Harpy		13″	1,200	160
35	Vizier		13″	1,250	160
36	Vandal		13″	1,200	160
37	Siren		13″	1,270	160
38	Sultan		13″	1,210	150
39	Wolf	2 × 5′0″ & 4 × 3′6″	14″	1,100	150
40	Shark	5′0″ CPLD.	12″	1,300	160
41	Ganymede	4 × 5′0″ CPLD.	12″	700	150
42	Buffalo	4 × 5′0″ CPLD.	13″	1,300	160
43	Bloodhound	4 × 5′0″ CPLD.	13″	1,300	160
44	Mastiff	4 × 5′0″ CPLD.	13″	1,350	160
45	Mammoth	4 × 5′0″ CPLD.	13″	1,200	160
46	Fox	4 × 5′0″ & 2 × 3′0″	13″	1,375	155
47	Rob Roy		13″	700	90
C	spare tender				140
D	spare tender				100
				£47,800	£6,640

TABLE 10   *The Locomotives*

principle, if it be deemed desirable by the Railway Company.''

ORDERS AND DELIVERIES

In reply to the Butterley Company's letter, on 13 March, 1839, the North of Trent Works Committee resolved that two engines, one on four and the other on six wheels, be ordered on the terms proposed and that the proposal to contract for the maintenance and working on the plan adopted by the London and Birmingham Railway be deferred for future consideration.[2] At the same time, the Lancashire Directors were requested to advise upon the description of the locomotives and of the first and second class carriages required, with suggestions as to suitable builders.[3]

Theodore Rathbone, Edward Cropper and Robert Garnett, already experienced in railway matters, championed Edward Bury's inside cylindered four-wheeled design of engine, as used on the London & Birmingham Railway. As a result of Rathbone's strongly expressed views the board ordered six engines of the Bury type, three each from Jones, Turner & Evans of Newton, Lancashire, and Stark & Fulton of Glasgow, at a cost of 1,125 and 1,000 respectively plus 150 each for tenders. The propensity of inside cylindered engines to suffer from breakage of the crank axles, however, had led Vignoles, the company's engineer, and his assistant, Thomas Woodhouse, to support the use of the outside cylindered engines ordered from Butterley. A further 23 engines of the Bury type were ordered in January 1839, some being delivered with the larger 13in diameter cylinders which became an obvious requirement on the opening of the line, since the 12in cylinder engines exhibited a distinct shortage of power. More followed later, so that of the total fleet of 47 engines passed over to the Midland Railway, only seven were not of the Bury type or built by his company.[4] Of the eight engines used by the contractors only one, or possibly two, were absorbed into the company's stock.[5]

As mentioned elsewhere, Stark & Fulton, the Scottish builders of the first engine to be received (No.3 *Hawk* in the attached list), were somewhat inexperienced and delivered the machine in an unfinished state during April 1839. It was followed by three engines from Jones, Turner, and Evans, No.4 *Sunbeam* on 18 April, No.5 *Wizard* on 5 May, and No.6 *Hecate* ''last Tuesday'' (relative to 17 June), these being put into service on 20 April, 18 May and ''this week'' (as above). No.7 *Lion* was delivered by Edward Bury about the same time. By 10 December 1839 eight engines, eventually numbered 3–10, were in service, No.11 following in January 1840. No mention of engine numbers was made in the report on the opening of the line between Nottingham and Derby on 29 May 1839 and it is possible that the engines at first carried only their names; they are supposed to have been numbered by October 1839. The Butterley engines did not perform very well and payment was disputed. Although they

TABLE II   *Moveable Tools in Use*

Smith's shop	334.	0.	0
Fitting up shop	327.	0.	0
Bolton turning shop	164.	0.	0
Tinmans tools	3.	15.	2
Copper smiths tools	35.	19.	8
Casting shop	13.	10.	7
Boiler shed	78.	10.	0
Painters shop	4.	4.	6
Shop in shed	5.	11.	5
Engine shed	54.	10.	0
In the yard	35.	4.	0
Upper turning shop	194.	15.	0
Brass finishers shop	13.	5.	0
End of coal shed	32.	0.	0
Joiners shop	10.	4.	8
Coke store	5.	18.	6
Cobblers tools	1.	10.	0
40 toolboxes and engine drivers tools	260.	0.	0
Leather belts for lathes	30.	0.	0
Articles in store room	37.	0.	0
Tools at Nottingham	81.	10.	0
Do. at Rugby & Leicester	150.	0.	0
	£1,872.	8s.	6d

were delivered in May 1839 and April 1840 it was not until November 1841 that the Finance Committee paid 3,447 for them.[6]

Alterations to the stock had been made by the time, on 13 February 1843, that John Hick of Bolton carried out his valuation, and it is from this document that the details set out in Tables 10 and 11 have been obtained. *Ariel*, one of the engines mentioned in the report on the opening ceremony on 29 May 1839, seems to have been renamed *Bee* in 1841, while No.40 *Harlequin* was renamed *Shark* in 1842–43. Engine No.42 *Buffalo* was furnished with an iron firebox and tubes.

It is only partially clear, from the details given in Hick's valuation, that the wheel notation of engines nos.1 and 3–38 was of the 2-2-0 pattern. Nos.2, 39 and 47 were 2-2-2s, Nos.41–45 0-4-0s, and No.46 an 0-4-2. Outside cylinders were provided on Nos.1, 2, 16–19, and 32–38, the stroke of all engines except Nos.1 (16in) and 46 (20in) being 18in. On the Midland Railway, where the stock was grouped according to wheel type and class, the 47 Midland Counties engines seem at first to have been given the numbers 136, 23, 111–123, 100–103, 124–135, 104–110, 21, 83, 84, 79–82, 85, and 22. Two long boilered 0-6-0s were ordered from B. Hick & Son of Bolton in December 1843 but were not delivered until after the amalgamation, paid for by the Midland Railway and taken into its stock as Nos.75 and 76.

Deliveries appear to have been made to the following sequence: Nos.3 and 4 in April 1839, Nos.1 and 5 in May 1839, Nos.6 and 7 in June 1839, No.47 in July 1839, Nos.8, 9 and 10 in September 1839, No.11 in January 1840, Nos.20, 21 and 22 in March 1840, Nos.2, 23, 24 and 25 in April 1840, Nos.12, 16, and 17 in May 1840, Nos.19 and 26 in June 1840, Nos.13, 27, and 28 in July 1840, Nos.14, 15, 18, 39, 40 and 41 in August 1840, Nos.29 and 31 in September 1840, Nos.30 and 46 in October 1840, Nos.42, 43 and 44 in November 1840, Nos.35 and 36 in April 1841, Nos.32, 37 and 38 in May 1841, No.45 in June 1841, and Nos.33 and 34 in August 1841. No.46 *Fox*, new in May 1840, had been delivered to

Eckersley & Worswick, contractors on the section between Loughborough and Syston, being taken into Midland Counties Railway stock as indicated above. According to Barnes, No.47 *Rob Roy* had also been used by one of the contractors, together with *Mersey*, (built by Galloway of Manchester), *Aphrite, Etna, Navy, Trent,* and *Vivid.*

Seven manufacturers were involved in supplying the locomotives. Nos.7, 8, 11–19, 35–38, and 42–45 were built by E. Bury & Co., Nos.1 and 2 by the Butterley Company, Nos.40 and 41 by Wm Fairbairn & Sons, Nos.26–34 by B. Hick & Son, Nos.4, 5, 6, 45 and 47 by Jones, Turner & Evans, Nos.20–25 and 39 by Nasmyth, Gaskell & Co., and Nos.3, 9, and 10 by Stark & Fulton. All of the 2-2-0s (except No.1 from Butterley) were built to Bury's design, some with inside and some with outside cylinders as indicated above.[7]

SUPERINTENDENCE

Josiah Kearsley had been appointed locomotive superintendent for three years in January 1839 at a salary of £20, rising by £50 per annum, and he held the post throughout the company's independent existence.[8] At amalgamation he became one of three contenders for the single new post of superintendent of the Locomotive and Carriage Departments. The others were Thomas Kirtley of the North Midland Railway and his younger brother Matthew, in charge of the comparatively smaller establishment of the Birmingham & Derby Junction Railway. Born at Tanfield, County Durham, in 1813, Matthew started his working life as a tailor but soon forsook that trade for locomotive engineering and joined the Warrington & Newton Railway. Thence he moved to the Leeds & Selby (from which he was dismissed) and the London & Birmingham concerns. He entered the service of the Birmingham & Derby Junction Railway in 1839 as locomotive foreman at Hampton and, at the early age of 28, became Superintendent, largely, perhaps, because of his parsimony in the matter of wages, which effected substantial reductions and pleased the board when retrenchment became necessary.

The reasons for Matthew Kirtley's appointment as Locomotive Superintendent of the Midland Railway on its formation are difficult to understand. By comparison with his brother and more particularly Kearsley he was ill-trained and well below the standards of the period.[9] Yet, thanks no doubt to strong recommendations by the Stephensons, under whom he had served as a pupil, it was young Matthew who was appointed, at the age of 31, to the post which he retained until his death in 1873. His commencing salary was £250 per annum, some £50 more than his previous earnings. Brother Thomas accepted the choice, staying on with the Midland Railway for a time and working under Matthew as an inspector from March until May 1845, when he was paid £100 and went to work for Thomas Brassey on the Trent Valley Railway. Kearsley, who fully expected to be offered the position, became very embittered. Although he still had 2½ years of his renewed contract to perform, he refused to serve under Kirtley and removed his affections to Messrs Rothwell & Co, locomotive builders, of Union Foundry, Bolton-le-Moors.[10]

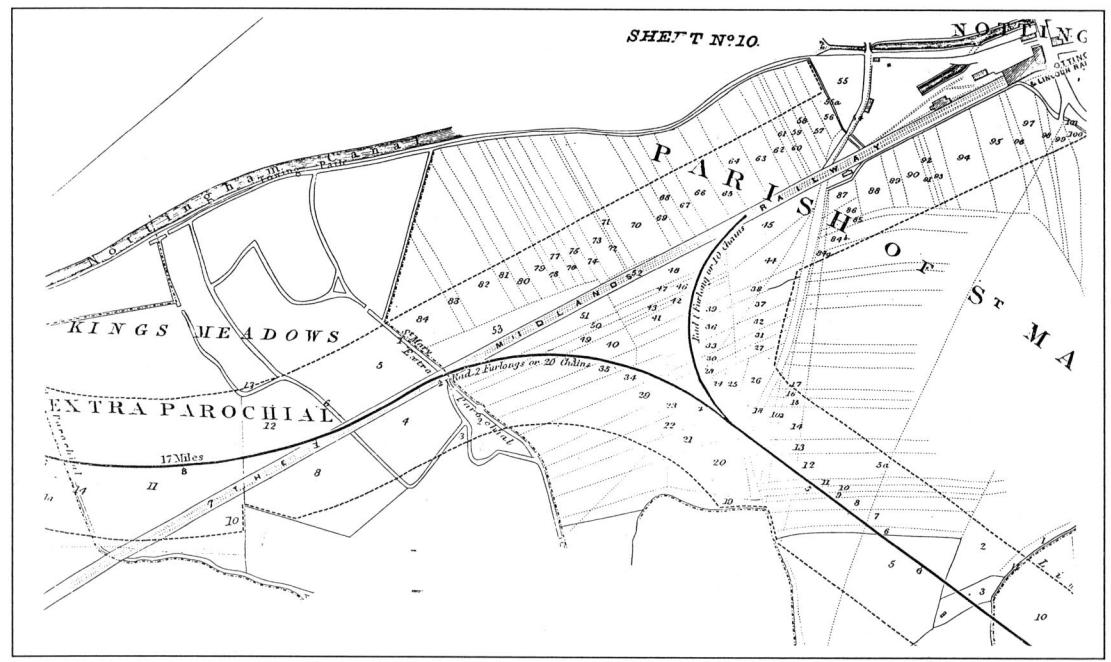

38. The first Bill for construction of a railway between Ambergate and Boston planned junctions with the Midland Railway system at Lenton and at Carlton, with an alternative route around the station through West Bridgford. Deposited a year later than those for the Lincoln line, the plans give even less detail of the Midland Counties terminus, already threatened by the alterations consequent upon the construction of this extension. Nottinghamshire Records Office.

39. In order to cater for the Lincoln traffic it was necessary to graft onto the Midland Counties premises additional platforms in a new single gabled train shed. This rather poorly reproduced illustration seems to form the only surviving record of the extended premises some time after they had been given up in favour of the new establishment in Station Street. By the end of 1864, when plans were being made for further relief lines around the old station and the raising of Carrington Street and the adjoining roads onto a viaduct over the railway, there were four sets of rails across the road, two emerging from the Lincoln train shed and two passing outside. Nottingham Historical Film Unit.

7

Epilogue

It was inevitable that Derby should become the headquarters of the new company, since the geographical centre of the enlarged system had obviously moved away from Leicester, where the Midland Counties Railway had established its principal offices. Hudson was the obvious choice for chairman, with John Ellis, formerly a director of the Midland Counties as deputy; in May 1849, when Hudson was forced to retire, Ellis succeeded until he in turn was replaced by George Byng Paget, another of the Midland Counties men. John Fox Bell, secretary and superintendent of the Midland Counties Railway, served as secretary to the Midland Railway until 1853 and William Henry Barlow, who had succeeded Woodhouse as resident engineer of the Midland Counties remained as chief engineer until 1857, after which he acted in a consultancy role to the new company. The other directors who stayed on were Lawrence Heyworth, William Evans Hutchinson, John Taylor, Samuel Waters and Henry Youle.

Before the amalgamation took effect it was clear that the company's premises in Nottingham would have to be radically altered to cater for the Lincoln traffic. The booking hall and office building blocked any exit from the train shed in an easterly direction and it was obvious that, short of demolishing this frontage, the Lincoln line must make a junction with the existing railway at some point west of the station. Additional platforms were to be provided on the through lines to the south of the existing train shed in order to obviate reversal of the traffic to and from the original terminal platforms. Whether these additional platforms were available when traffic began on 4 August 1846 is not clear but they were certainly in place before the end of the year.[1]

By the time that the Lincoln line had been completed, such a short time after the passing of the relevant Act in June 1845, it had become clear that Nottingham station was likely to have to cater for traffic from several other directions. A nominally independent company having obtained its Act,[2] the Midland Railway company had subsequently taken over their powers as agreed and was already busy with the construction of the Erewash Valley Railway to meet the Derby and London lines south of Long Eaton. On 16 July 1846 Acts were passed sanctioning construction of a Midland Railway branch down the Leen Valley from Mansfield, which would join the Derby line at Lenton and for a truly independent venture grandiosely styled *The Ambergate, Nottingham, and Boston, and Eastern Junction Railway*.[3] Intended as just one link in a chain of new railways stretching from Manchester into the eastern counties, this line had been planned to run from a junction with the North Midland Railway at Ambergate through Codnor Park in the Erewash Valley and parallel to the Nottingham Canal towards Radford. From this point its route lay so close to that of the Midland Railway's projected line from Mansfield that Parliament decreed joint construction as far as the intended junction with

40. The extension to Lincoln

W. Deardon's prospect of Nottingham, published in January 1850, shows trains simultaneously arriving from Long Eaton and departing towards Lincoln. Representation of the old and new station premises is not, however, very well handled. Brewhouse Yard Museum.

the Midland Counties line at Lenton.

No agreement having been reached for use of the Midland Railway company's lines and station, the "Ambergate" railway had been planned to circle southwards through West Bridgford as an alternative to the use of the Lincoln branch metals as far as Carlton, where the newcomer once again diverged through Ratcliffe, Bingham, Bottesford, and Grantham on its way towards Sleaford, Spalding, and Boston. With the prospect of all of this additional traffic the Midland Railway company moved its station to a more commodious site eastwards along the Lincoln line into the West Croft, where, having obtained an enclosure Act in 1839, the council had constructed a branch out of the Nottingham Canal. Between this canal, London Road, and Station Street the company laid out a bigger and even more elegant station than that of the Midland Counties Railway,

removing its traffic into these new premises on 26 May 1848. Behind the classical offices facing into Station Street there were a range of train sheds symmetrically disposed about the through Lincoln lines and an open turntabled semi-roundhouse for locomotives.[4]

These were the premises into which the "Ambergate" services from a terminus by the canal wharf in Grantham commenced to run on 15 July 1850. Although the company had contributed its fifty per cent share towards the cost of the southern end of the Midland Railway's Mansfield branch this was reimbursed when it became clear that, due to shortage of funds, the remainder of the line from Ambergate would not be made, along with all sections east of Grantham.[5] The name of the company was later changed to give effect to these reduced ambitions; from 15 May 1860 it became known as the Nottingham & Grantham Railway & Canal

41. T. Stevenson's engraving of the exterior of the new station in Station Street, Nottingham. Expansion of the railway system and consequently increasing traffic demands had quickly made the Midland Counties establishment inadequate for its purpose. Nottinghamshire County Libraries.

42. S.C. Hodgson's engraving faithfully captures the spacious interior of the new station at Nottingham, set symmetrically astride the through lines to Lincoln. Although at least four years after amalgamation the Midland Counties Railway initials still feature prominently upon the smokebox of the locomotive. Nottinghamshire County Libraries.

43. M. Webster's view of the new station in the West Croft, looking along Station Street towards the original premises. The bridge across the canal in Carrington Street, paid for in large measure by the railway company, is particularly noticeable. Brewhouse Yard Museum.

Company, duly acknowledging the enforcement of a pledge to purchase the Nottingham and Grantham Canals. Shortly afterwards the arrangement by which the Great Northern Railway was already working the line was turned into a long-term lease. Before this association had been so well cemented, however, upon the opening of the Great Northern Railway line through Grantham in 1852 and the arrival in Nottingham of through services from Kings Cross as well as from Euston, still via Leicester and Rugby, relationships between the companies deteriorated. The Midland company impounded a Great Northern locomotive, supposedly leased to the "Ambergate" concern, and threatened to throw that company's booking clerk and his tickets out of the booking office.[6] Closer ties with the Great Northern company, despite an injunction obtained by the Midland Railway against the first proposal for a lease, and an independent line into Nottingham from Colwick, opened on 3 October 1857, removed much of the pressure from the Midland

station. Like the one at Leicester, it was extensively rebuilt around the turn of the century, expanding across all of the land previously occupied by the West Croft canal and westwards to a new entrance on the Carrington Street viaduct.

Near and far, down the years, many changes were made to the ever expanding Midland Railway system which impacted upon the pattern of services over the former Midland Counties lines; the changes at Trent Junction have already been mentioned. Additional traffic was brought onto the system with the opening of the Syston and Peterborough line (1848) and of the Leicester and Burton railway (1849), the latter being an extension of the Leicester & Swannington system, now incorporated into the Midland empire. Reversal at Derby of all through services, to and from the former North Midland route, was rendered unnecessary in June 1867, when the Litchurch Curve was opened between Spondon and London Road Junctions. Trains could then be brought off the former Midland Counties lines into Derby Station either from the north, over the original route through Chaddesden, or from the south, over the new line, proceeding directly onwards towards Birmingham or Leeds as required. Additional wayside stations were provided at Draycott (July 1852), Attenborough (September 1864), Hathern (February 1868), and Sawley Junction (December 1888). The original establishment at Borrowash was replaced by a new one lying ¼ mile farther west in January 1871.

Alternative routes were provided; from Trent (Sheet Stores Junction) through Weston-on-Trent (December 1869) onto the Birmingham and Derby line at Stenson (November 1873), so allowing south-west bound goods traffic to avoid Derby; from Radford, on the Mansfield line, to Trowell, on the Erewash Valley line, in May 1875, facilitating traffic between Nottingham and Manchester via Codnor Park and Ambergate instead of through Derby; and, again for goods traffic, the overhead lines from the Toton marshalling yards in the Erewash Valley to Attenborough, Trent and, over a new bridge and through a second tunnel at Redhill, to Radcliffe on Soar in May 1902. From March 1880, with the opening of the lines between Kettering and Manton, and between Melton Mowbray and London Road Junction, Nottingham obtained an alternative

44. Site plan of Nottingham New Station

Parliamentary plans, lodged by the "Ambergate" company in 1853 for making its station in London Road, show in some detail the new premises which the Midland Railway had erected between Station Street and the West Croft Canal in Nottingham. Turnplates out of the platform lines, either side of the through running lines to Lincoln, may be noted as well as the segmental open roundhouse for locomotives. Nottinghamshire Records Office.

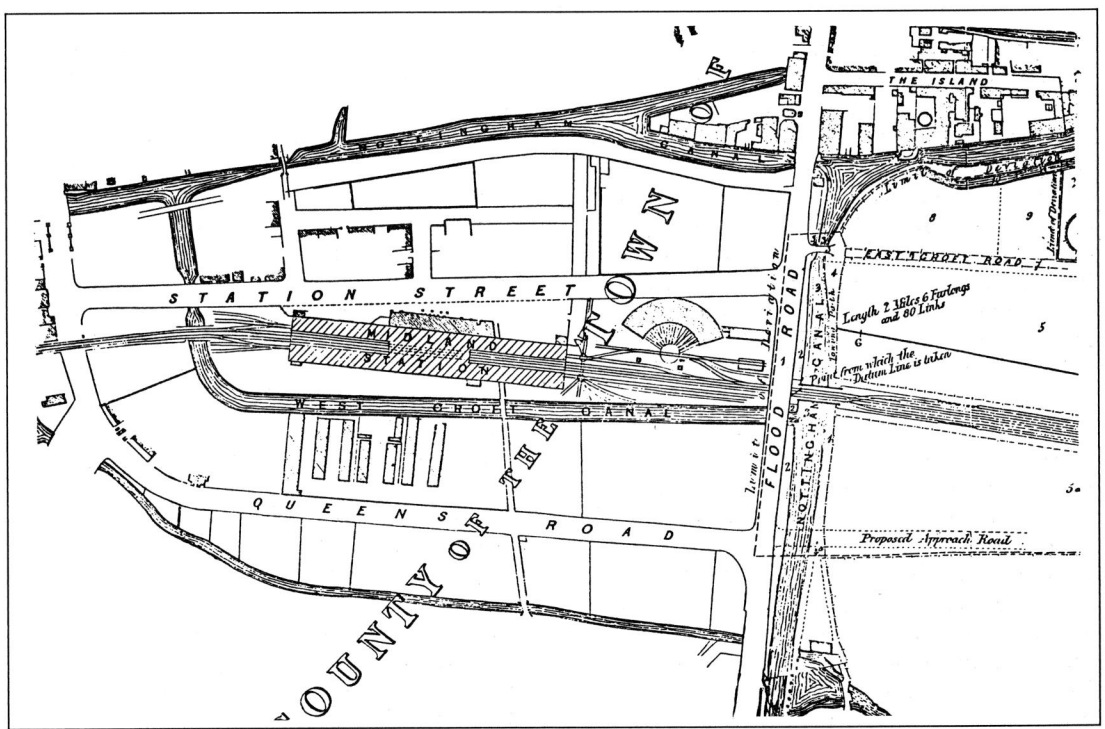

45. This view, by Johnson and Prior, shows excavations under way in preparation for the building of the locomotive depot at the new premises in Station Street. The platform shelters of the new station appear to have been finished and a siding made across the end of the excavation to accommodate spare carriages. Brewhouse Yard Museum.

route for its London traffic to those available through Leicester.

It was to the south of Leicester that the greatest changes took place. Euston having become overcrowded with all of the traffic funnelling into the west coast main line, and the Great Northern Railway being willing to receive more traffic into its terminus at Kings Cross, the Midland Railway Company decided to build a new line from Leicester (Wigston Junction) through Kettering, Wellingborough, and Bedford, to Hitchin, which was opened in May 1857. Although traffic was at first worked forward from Hitchin by the Great Northern Railway, from February 1858 Midland Railway trains ran into Kings Cross. In a few years, however, congestion on the Great Northern line prevented efficient operation of the timetable and the Midland Railway at last determined upon having its own London terminus, using the line from Bedford into St. Pancras for all traffic from October 1868. Bereft of its through traffic, the Midland Counties line south of Wigston continued in use for local traffic until the beginning of January 1962. Modified through the Trent area as early as 1862, and latterly shorn of the section through Chaddesden, the rest of the Midland Counties route remains fully used by Inter-City 125 trains and other traffic.

That the line should have survived for 150 years is a tribute to the soundness of the original promotion and of George Hudson's efforts to weld together the component elements of the Midland Railway. In view of their interest, the words of a contemporary economist, writing in February 1838, have been reproduced in Appendix 4. Before either of the lines was open, he foresaw that the Birmingham & Derby Junction and Midland Counties companies would be brought into a fierce competition for the through traffic from Derby to London. Success would depend upon the soundness of the local traffic, as well, in the case of the Birmingham & Derby Junction Railway, as that offering between the north-eastern and south-western parts of the country.

The 1832 Prospectus [1]

At a Meeting of Subscribers to the RAILWAY from PINXTON to LEICESTER, held this 4th day of October, 1832, EDWARD MILLER MUNDY, ESQUIRE, In the Chair; It was RESOLVED,

That a Provisional Committee be formed to superintend the affairs of the proposed Railway, and that the Members of the late Committee of Proprietors and Lessees of Collieries of the Counties of Derby and Nottingham, being Subscribers thereto, be requested to continue their services, until Directors and other Officers are appointed, with authority, in the mean time, to add to their number, and to take such preliminary measures as may be necessary for disseminating a knowledge of the great public advantages involved in the undertaking, as well as for obtaining the sanction of Parliament for the Construction of the proposed Railway at the earliest possible period.

E.M. MUNDY, Chairman.

Alfreton, October 15th, 1832.

The Committee appointed by the foregoing Resolution, in compliance with the wish therein expressed, have prepared the following general explanation of the origin, objects, and advantages of the Railway from Pinxton to Leicester, with its contemplated extensions to Nottingham and Derby, which they propose to denominate the MIDLAND COUNTIES RAILWAY, and trust that the facts and observations they have subjoined will be conducive to the end desired, of disseminating a knowledge of the great public benefits which are contingent on its execution.

MIDLAND COUNTIES RAILWAY

The construction of a Railway from Leicester to Swannington, and the speculations in progress for bringing the coal of the contiguous district into the Leicester market, having threatened the collieries of Derbyshire and Nottinghamshire with the loss of that portion of their trade which they have hitherto enjoyed along the navigation of the Soar – amounting to a quantity, perhaps, not less than one hundred and sixty thousand tons annually – and the same causes tending, in their consequences, to deprive the navigations, connected therewith, of their revenue arising from this source, the coal proprietors of the counties last mentioned, with a desire to avert so serious a common misfortune, endeavoured to induce the navigation companies to make such reduction in their rates of tonnage, as, with some sacrifice of price on the part of the collieries, should retain the trade, or, at least, a portion of it, in its ancient channels.

After many fruitless attempts, in the course of which the Leicester and Swannington Railway derived the support necessary for its completion, the coal proprietors, warned of the near approach of the danger, as it regarded themselves, adopted the only alternative open to them, by proposing the formation of a Railway to Leicester, and inviting the co-operation of the public in its execution, as a mode of profitable investment of the capital it would require.

Since this alternative was determined upon, three of the navigation companies, above adverted to, have reduced their tonnages sixpence per ton each; but the relief thus afforded will not enable the coal proprietors to maintain their position in the Leicester market, the expense they have yet to incur, in tonnage alone, being 3s.2d. per ton from the nearer collieries, and 4s. per ton from those farther off, to which is to be added a charge of 2s.6d. per ton for freight, making a total of 5s.8d. per ton in the one case, and 6s.6d. per ton in the other, for the cost of conveyance to Leicester, whilst the Leicestershire collieries, at present, incur an expense of 2s.6d. per ton only, in delivering their coals at the same point; and it has, recently, been stated, on the authority of one of the largest shareholders in the Leicester and Swannington Railway, that, provided the traffic increase to the extent expected, this charge may be reduced to 2s., and even to 1s.6d. per ton, with a return of 5 per cent. on their capital to the shareholders!

The authentication of this fact must be admitted as a complete and final answer to the question, whether the navigations are capable of affording the collieries the assistance necessary to their future participation in the coal trade of Leicestershire, and as it at the same time assures the certain loss

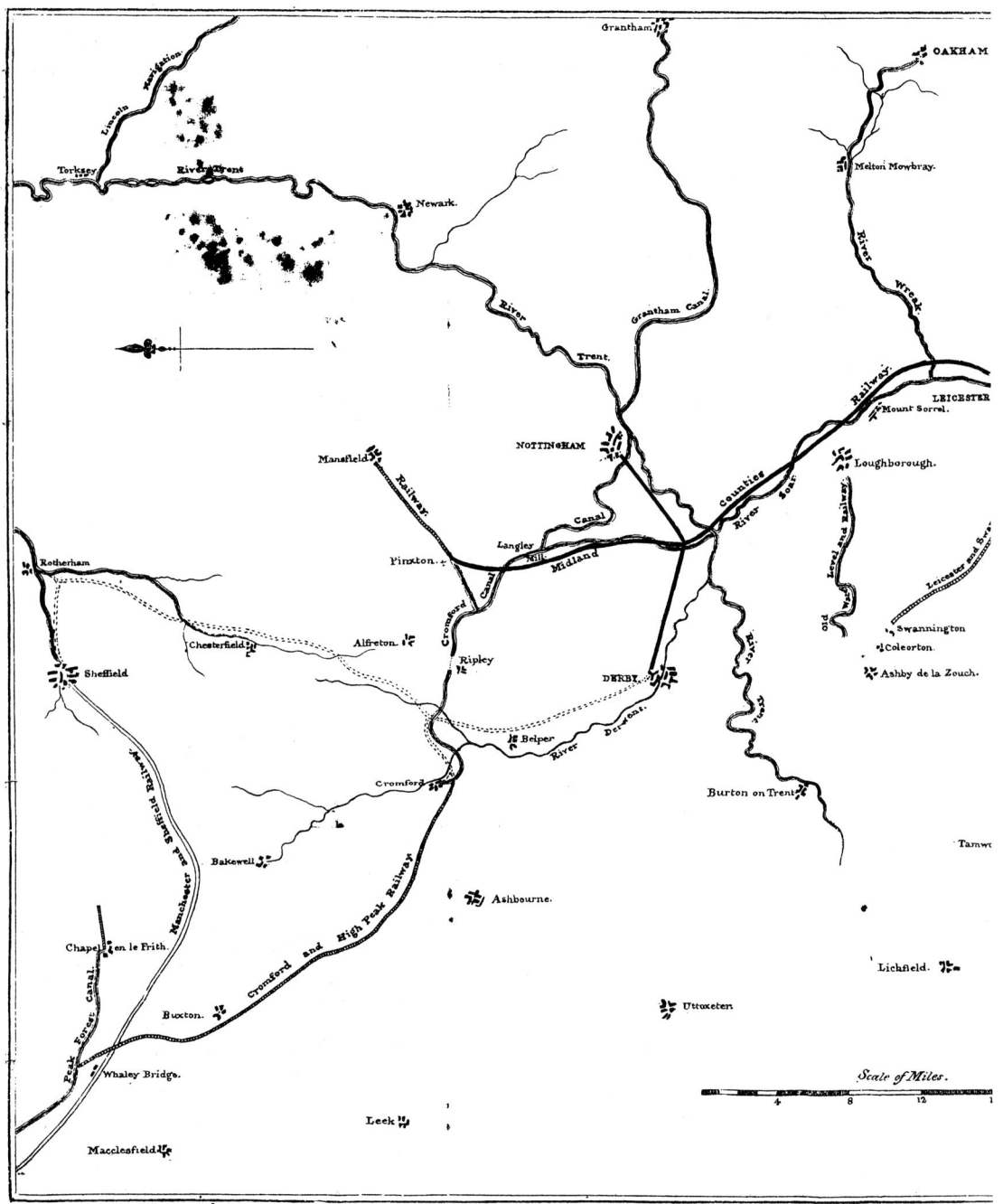

46. The map accompanying the original prospectus of the Midland Counties Railway indicated alternative routes south of Leicester connecting with the London & Birmingham Railway at Rugby and south of Northampton. Original courtesy of Leslie Hales.

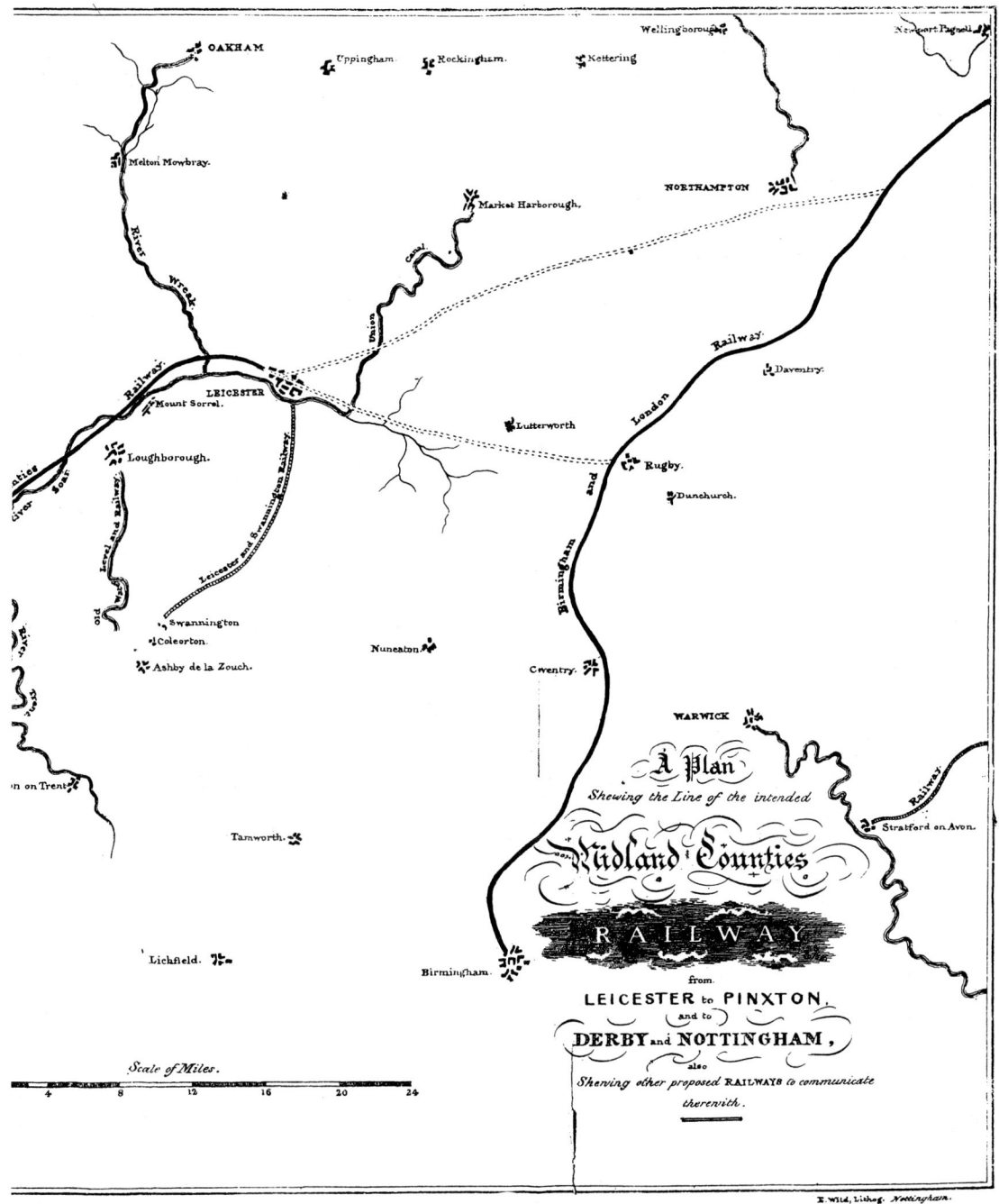

of this traffic to the navigations, whether any other mode of conveyance be provided, or not, the promoters of a Railway are thus exempted from all imputation of rendering the capital vested in the navigations unproductive, since, it is evident, that, if the trade arising from coals be not transferred to a Railway, it must and will be lost to these Counties altogether, thus injuring, in addition to the navigations, the owners of all coal property with which they are connected, together with a very extensive population employed in the collieries; and all who, either directly, or indirectly are concerned in administering to their wants. The interests involved in this last consideration are so important and extensive, that every landholder in the neighbourhood of the coal mines becomes personally identified with the success of the efforts now making to preserve them in their accustomed state of active employment.

To what extent, a Railway, to be formed for these objects, is capable of recommending itself as a source of profit to subscribers, must depend, conjointly, on its cost, and general utility. With reference to the former, the Committee appointed to the present duty, having had access to the very elaborate and complete surveys and estimates of the late Josias Jessop, Esq., made for the intended portion of the Northern Railway from Pinxton to Leicester, have assured themselves that a double line of Railway, between these points, passing along the vallies of the Erewash and Soar, and constructed on the most perfect system adapted to locomotive power, will cost £200,000; and they have assumed, as they believed, without much risk of error, that a single line may be completed for about one third less, or £130,000. The natural advantages presented in its proposed course are almost unequalled, affording to the Road an uniform fall to the Trent, in the direction of its principal trade, of twelve feet in a mile, and a rise in each mile, from thence to Leicester, of four feet, the greatest embankment in its whole length, with two or three inconsiderable exceptions, being only about ten feet in height!

In enumerating the sources from which its income is to be derived, the Committee, firstly, advert to the staple article of coal, of which they assume that 160,000 tons may annually be vended by the collieries, as heretofore, in the markets south of the river Trent; for, although their new competitors in Leicestershire will, undoubtedly, absorb some portion of the demand, – possibly, the half of what, at present, exists, – it is, at least, as certain that the reduction of not less than 4s.6d.

per ton, in the price of coals, consequent on the opening of the Railway, will add very greatly to the range of market, hitherto, enjoyed by these collieries, especially in the direction of the Wreak River, eastward of Melton Mowbray and Oakham, where the coals of each district will meet at equal distances, but, extending, also, through the whole of Rutland, Northamptonshire, and Buckinghamshire, and to a great part of the contiguous counties of Huntingdon, Cambridge, and Bedford, the supplies for which are now drawn from sources that afford no adequate inducement for applying to them like improved facilities of conveyance.

As auxiliary to this increased sale, the superior quality of the coal produced in Derbyshire and Nottinghamshire is entitled to a particular notice, *being already valued by consumers* at from one shilling to two shillings per ton more than the best coals yet obtained from the new collieries of Leicestershire, by the price they consent to give for them in the same market. With regard to the cost of conveyance to which the former coal proprietors will be exposed, they have the authority of Mr. Stephenson, who has been described to them as "one of the most eminent Engineers this country has ever known," for stating its utmost limit at one third of a penny per ton per mile; the printed Report to the Sheffield and Manchester Railway Company, proceeding from the same authority, having recorded this as the cost, "including the expenses of engine power, and *attendants* upon the *coaches* and waggons," on a double line of Railway between Manchester and Stockport, rising seven feet in each mile, and moving 150,000 tons per year (page 19). On another division of the same road, with an inclination of thirty four feet in each mile, the same expenses are estimated at one farthing per ton per mile (page 20). Deducting from these, the charge now incurred by the coal masters of Derbyshire and Nottinghamshire, in stocking their coals at their canal wharfs, and delivering them to the boats, which will not attach to a trade by Railway, the effective cost of conveyance from the collieries to Leicester, beyond their present payments, will not exceed sixpence per ton for the whole distance. A farther tax of one shilling and threepence per ton for the use of the Railway, being one halfpenny per ton per mile on the average distance of thirty miles, would complete the delivery of the coals in Leicester, from the collieries, for one shilling and ninepence per ton, and that such a charge would remunerate the proprietors of the Railway, in addition to the other branches of income to arise upon it, will be found, in the farther progress of these observations, scarcely to admit of doubt. The Committee, therefore, have the fullest confidence in enumerating the article of coal as constituting a permanent and principal source of traffic. Much more will undoubtedly arise from the lime and gritstone of Derbyshire passing southward, and from the granite of Leicestershire introduced for the public roads of these counties, where its use, from the great reduction of price at which it can be thus furnished, will become almost universal. Merchandise of all kinds and corn will be carried along the road in large quantities; and when it is known, with respect to lime, that the necessary and liberal use of this material in agriculture is now almost wholly suspended, from the expense of carriage, in the more distant parts of the country where the quality of the land so imperatively requires its application, a considerable source of revenue may be thence expected. In years of greater prosperity to the agriculturist, a quantity equal to forty thousand quarters of lime has been sold annually from one establishment alone on the Cromford Canal into Nottinghamshire and Leicestershire, which demand has now wholly ceased, although the lime works of the neighbourhood of Crich, which are employed exclusively in supplying the local consumption of the district, continue in unabated activity. These facts plainly indicate, not only the privation and loss which the more distant farmer is exposed to from the want of this important ingredient in his operations, but justifies the inference, that, when the carriage of it is reduced, and the cost again bears the same relation to the value of agricultural produce, as before, the former demand, will, at least, revive.

In the conveyance of passengers, and of the light packages transmitted by coach, the Midland Counties Railway is destined to be of the most extensive utility, benefitting alike the public, and its proprietors. The income estimated to result from these sources on the proposed Railway from Sheffield to Manchester, a distance of 43½ miles, was £24,200 per annum, and admitting the accuracy of the calculation from which that estimate was derived, a return, not less proportionably ample, may be anticipated from a line of road, which, by its direct and advantageous means of communication, cannot fail to attract so large a share of the intercourse between the northern and southern parts of the kingdom. By its aid alone, the distance between Leicester and Sheffield, and, consequently, between London and Sheffield, will

be, actually, shortened about seven miles, and, from the increased speed of Railway travelling, as compared with the usual modes of conveyance, these towns will thus, virtually, be brought nearer by a space equivalent to nearly three hours of time. Similar and equal advantages will be obtained by Mansfield, Alfreton, Chesterfield, Barnsley, and Leeds, and, in fact, by all towns northward of the latter, on the line of road with which it is connected, thus giving to the numerous and active population in their vicinities, including the greater part of the northern counties of England, and the whole of Scotland, a position, as measured by the usual rate of travelling, from twenty five to thirty miles nearer to the metropolis: But, as regards the intercourse by post, the consequences available from this undertaking are of still greater value and importance. The transmission of the London Mail for Sheffield, and certain of the intermediate towns, down this Railway, would insure its arrival in Alfreton *by half-past eight o'clock* each morning, instead of *after twelve o'clock,* as, at present; and the distance of Alfreton from Sheffield being only twenty-two miles, the delivery of the Mail, at the Post Office of Sheffield, could be effected each day by *eleven o'clock!* An interval of four hours being allowed in Sheffield for the delivery and making up of the letter bags, and the answering of letters, the Mail might be returned by the same route time enough to meet the Manchester Mail at Leicester the same evening, thus giving to Sheffield and the intermediate towns of Alfreton, Mansfield, and Chesterfield, with their contiguous Manufacturing districts, the very valuable privilege of a *daily exchange of correspondence with London,* all which would be effected at a *saving* to the Government of *two-thirds* of the *expense now incurred* in transmitting the Mail by the ordinary mode of conveyance, as has been fully established by the experience of the Liverpool and Manchester Railway.

To enable the public to derive the full benefit of these improvements, the Committee are not unmindful that a double line of Railway will be necessary, and that the capital requisite for its completion will be £200,000; but, they do not regard this as an objection or difficulty, nor can they doubt that the accomplishment of an object by which the benefits of internal intercourse will be so immensely facilitated, and so extensively diffused, even if purchased at much greater cost, could long be delayed. The towns of Nottingham and Derby, will very largely participate in these advantages, when the extensions contemplated to those towns are completed. The line connecting them will accompany the vallies of the Derwent and Trent, presenting every desirable advantage of surface, and being admirably adapted for the application of loco-motive power. The distance between these two towns will be only fifteen miles, and as the Railway will cross and unite with the line from Pinxton at Long Eaton, where the embankment over the valley of the Trent terminates, eight and a half miles distant from Derby, and six and a half miles from Nottingham, a communication with Leicester will be thus established, exceeding the present length of the turnpike road from Derby to Leicester only two miles, and from Nottingham to Leicester one mile. Each town will, by the same means, be placed on improved terms of communication with London for its merchandise; the distance by the Railway, and along the Union, Grand Union, and Grand Junction Canals, being less by nearly twenty miles than the present route, which, with the increased speed of Railway travelling, will save more than one day in the delivery of goods, and reduce the cost of carriage.

Such are a few of the advantages which the Midland Counties Railway presents as an appendage to existing establishments, and which its completion will, at once, confer; but, regarded as the commencement of a design, emanating, originally, from the spirited, and enlightened patrons of Railway communication in Liverpool, whose objects embrace the construction of certain principal lines of Railway throughout the kindom, whence branches might diverge to all parts requiring them, it rises immeasurably in importance, and executes its share of that great national enterprise. In the approaching Session of Parliament, the legislative sanction is confidently anticipated for the formation of a Railway from London to Birmingham, constituting another portion of the same great design, which passing, as it is intended to do, within a short distance of the town of Northampton, will, on the completion of the Midland Counties Railway, admit of a grand central communication being effected from London to Mansfield, by the extension of little more than thirty miles of Railway, southward from Leicester, to unite with the London and Birmingham line.

In conclusion, the Committee have only to add such further observations, relative to the pecuniary resources of the proposed Railway, as may suffice to illustrate and confirm the preceding statements. The average distance of the collieries from Leicester being thirty miles, a charge of one half-penny per ton per mile, on the assumed traffic of 160,000 tons of coal, will produce a revenue of £10,000 per annum. Adding to this, an equal sum *only* for passengers and merchandise, – and, comparing the intercourse and general trade along a principal thoroughfare like this with the same branches of income as expected to arise between Sheffield and Manchester, such a computation cannot be deemed excessive, – an aggregate income of £20,000 per year will be thence obtained for the use of the Railway, – waggons and moving power being provided by the carriers. In the Report on the Sheffield and Manchester Railway, already referred to, Mr. Stephenson estimates the cost of maintaining the road, for an annual traffic of 200,000 tons, at about £140 per mile (p.p. 19, 20). Proceeding on that calculation, and it is the result of actual experience, the cost of maintaining thirty four miles of the Midland Counties Railway would be under £5,000. A clear revenue of £15,000 for the use of the Subscribers would, therefore, remain, being seven and a half per cent. on the capital of £200,000 requisite for a double line of Railway.

Corrected surveys and estimates, adapted to the line now chosen, are in preparation, which, with an explanatory Report, will be submitted to the Subscribers at a meeting to be convened for that purpose, when it will be announced to them that responsible persons have presented themselves, who are willing to Contract for the completion of the work agreeably to these estimates.

Subscriptions are received at the Banking Houses of Messrs. WRIGHT, NOTTINGHAM; Messrs. CROMPTON and CO., DERBY; and Messrs. MANSFIELD and CO., LEICESTER; or at their respective London Correspondents, Messrs. ROBARTS, CURTIS, and CO.; Messrs. LEES, BRASSY, and CO.; and Messrs SMITH, PAYNE, and SMITHS.

Communications on the subject may be addressed to Messrs. LEESON and GELL, Solicitors, Nottingham.

William Jessop's Report [1]

TO THE *Subscribers to the Midland Counties Railway*
GENTLEMEN,
Having been requested by the Committee of Sub-scribers to the Midland Counties Railway to super-intend the revision and alteration of the Plans and Estimates for their proposed Railway from Pinxton to Leicester, the line of which had originally been laid down and prepared by my late brother, JOSIAS JESSOP, as part of the projected London and Northern Railway; I have proceeded to make such alterations in the line and in the levels as were required to adapt it to the object and views of its present promoters, availing myself of the experi-ence derived from the several public Railways established since its first projection.

I shall firstly describe the course the Railway pursues from Pinxton to Leicester, premising that the line is as direct as the great features or outline of the Vallies of the Erewash and Soar will admit, all other deviations from a straight direction being avoided, and the inflections being so slight at to admit of the utmost rapidity of motion which is consistent with safety, without danger of the displacement of the Carriages from the line of Railway.

The Railway will form a junction at its com-mencement at Pinxton with the one constructed some years ago from Mansfield to the former place, and proceed for three quarters of a mile by the side of a small river, the Erewash, which I propose shall be straightened, and in this part form the boundary of the Railway and of the adjoining property. Leaving the River on the right, the Railway will then cross the Alfreton and Nottingham Road on the same level as the latter, but in a situation, at the foot of Pye Bridge Hill, favourable for the erection of a Bridge over the Railway, if deemed necessary. It will then wind gently round the Hill, near Mr. Oakes' Selston Colliery, and, crossing the valley of the Erewash, at an elevation of about eight feet above the surface of the land, and afterwards the Cromford Canal at the foot of the Lock near the Codnor Park Ironworks, at a suitable elevation for navigation, it will proceed for some distance near to, and parallel with, the Canal, and in one straight course to the Turnpike Road between Heanor and Lang-ley Mill, near the lane leading to Milnhay. The

Cromford and Langley Mill Road will be passed in a cutting 15 feet deep, by an Archway over the Railway, and the Heanor Road being on the same level as the Railway, and not much used by the public, an Arch there will be unnecessary, and would be inconvenient. The line then proceeds through Lacey Fields, Shipley, and below Cotman-hay Wood, near to Mr. Thomas Potter's Farm House, and continues near the Erewash Canal through Ilkiston Parish, passing behind Mr. Samuel Potter's House. The Railway crosses the Erewash Canal near Gallows Inn, and afterwards recrosses for the purpose of keeping its direct course. Considerable excavations will be made at Cotmanhay and at Mill Field Lane, but they are required for raising the low parts of the Valley, and will supply only sufficient earth for that purpose. Further on the little stream called Nut Brook is crossed, as well as the Canal bearing that name, below the first Lock; this will be effected by a Bridge sufficiently elevated for navigation; these works are in the parish of Stanton. Sandiacre is then entered, and the Erewash Canal once more crossed at a sufficient elevation, being, in all cases not less than the Bridges constructed by the Canal Company. The parish of Stapleford is passed near to its boundary, and in this place the Railway crosses the Nottingham and Derby Turnpike Road, for which latter object, it is proposed to raise and make a deviation in the road, and to construct an Arch to conduct it over the Railway. The line is continued from thence through Toton and Long Eaton to the Trent, crossing the Nottingham and Sawley Bridge Road near Long Eaton. At the latter place, the embankment over the Valley of the Trent commences, which is, nearly uniformly, eight feet above the surface of the land, and is also above the level of the highest flood. The Cranfleet Cut, belonging to the Trent Navigation Company, will be crossed by a Bridge of a proper elevation; and for passing the River Trent, I propose to construct a Stone Bridge having six elliptical Arches, each of fifty feet span, for which the site is peculiarly favourable, having for the foundation the same strong red Marl and Gypsum Beds that are seen in the adjoining Red Cliff, through which the Railway will pass by a deep cutting, and thence supply the greater part

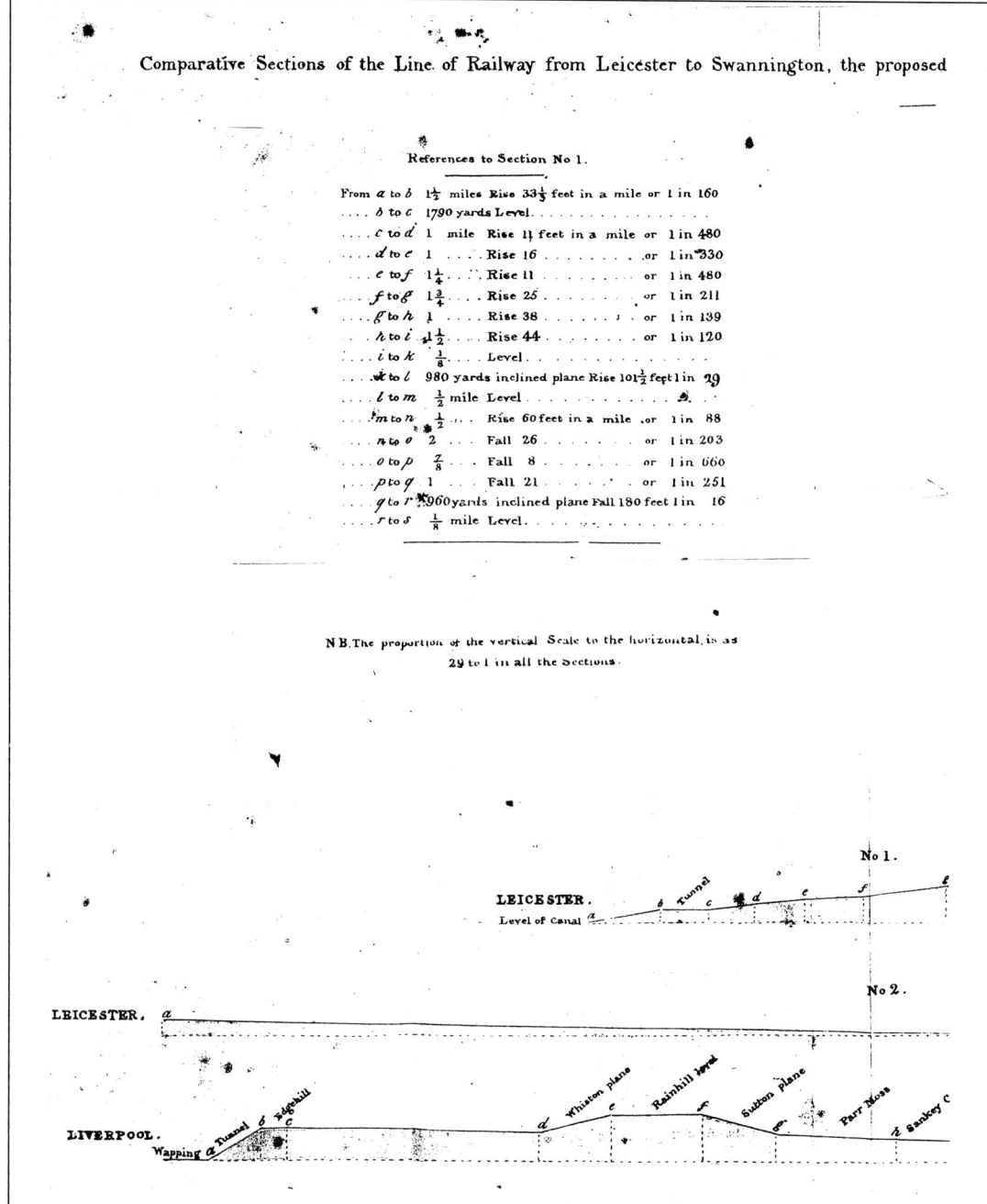

Comparative Sections of the Line of Railway from Leicester to Swannington, the proposed

References to Section No 1.

From a to b 1½ miles Rise 33⅓ feet in a mile or 1 in 160

.... b to c 1790 yards Level . . . .

.... c to d 1 mile Rise 1¼ feet in a mile or 1 in 480

.... d to e 1 . . . . Rise 16 . . . . .or 1 in 330

.... e to f 1¼ . . . . Rise 11 . . . . or 1 in 480

.... f to g 1¾ . . . . Rise 25 . . . . or 1 in 211

.... g to h 1 . . . . Rise 38 . . . . or 1 in 139

... h to i 1½ . . . . Rise 44 . . . . or 1 in 120

... i to k ⅛ . . . . Level . . . .

.... k to l 980 yards inclined plane Rise 101½ feet 1 in 29

.... l to m ½ mile Level . . . .

.... m to n ½ . . . . Rise 60 feet in a mile .or 1 in 88

.... n to o 2 . . . . Fall 26 . . . . or 1 in 203

.... o to p ⅞ . . . . Fall 8 . . . . or 1 in 660

, .. p to q 1 . . . . Fall 21 . . . . or 1 in 251

.... q to r 960 yards inclined plane Fall 180 feet 1 in 16

.... r to s ⅛ mile Level . . . .

N B. The proportion of the vertical Scale to the horizontal, is as 29 to 1 in all the Sections.

47. Comparative sections of the line from Leicester to Swannington, of the proposed Midland Counties from Leicester to Pinxton, and of the line from Manchester to Liverpool.

Line of the Midland Counties Railway from Leicester to Pinxton, and the Line from Manchester to
Liverpool,____

References to Section No 2.

From a to b $18\frac{3}{4}$ miles   Fall 4 feet in a mile or 1 in 1320
.... b to c $1\frac{2}{4}$ .... Level ....
.... c to d $11\frac{5}{8}$ ... Rise 12 feet in a mile or 1 in 440
.... d to e 2   ... Rise 14 .... or 1 in 377

References to Section No 3.

From a to b 1970 yards   Rise 123 feet or 1 in 48 (Inclined plane, worked by a fixed engine.)
.... b to c 1000 yards   Level ....
.... c to d $5\frac{1}{4}$ miles   Fall 5 feet in a mile or 1 in 1092
.... d to e $1\frac{1}{4}$ ... Rise 55 ..... or 1 in 96
    e to f $1\frac{7}{8}$ ... Level ......
.. f to g $1\frac{1}{2}$ ... Fall 55 ..... or 1 in 96
g to h $2\frac{1}{2}$ ... Fall 2 ..... or 1 in 2640
.. h to i $6\frac{1}{2}$ ... Fall 6 ..... or 1 in 880
... i to k $5\frac{1}{2}$ ... Rise $4\frac{4}{10}$ ..... or 1 in 1200 (Chat Moss)
... k to l $4\frac{1}{2}$ ... Level ... .

of the earth requisite for forming the embankment over the Trent Valley. The situation of the Trent Bridge will be on a part of the river not navigable, having its position immediately above the Weir, and below the confluence of the Soar and the Trent. Touching upon Thrumpston, the Railway passes through Ratcliffe, and Kingston, and below the Villages of Sutton St. Anne's, and Sutton St. Michael's, through the Parish of Normanton into Stanford, where the River Soar will be crossed by a Bridge at a part that is not used for navigation; then taking its course through Thorpe-acre into the Parish of Loughborough, at about one third of a mile from the Town, it again crosses the Soar in the Parish of Quorndon by a Bridge, under similar circumstances as the last, entering Barrow Parish and passing below the Town, beyond which a circuit of the River again intercepts the line of the Railway, and this I propose to obviate by diverting that portion of the River's course which will, also, shorten and improve its navigation. From Barrow the Railway proceeds into Sileby and Cossington, reaching the navigable river Wreak, and as, at this point, Wharfs will be required as a means of communication with the Railway from the district into which that navigation extends, the Railway is here brought down to a proper level for that object. It will, therefore, be necessary to pass this River by a Draw Bridge. The Railway then progresses through Syston, Barkby, and Thurmaston, and passing under the Foss Road by an Archway, in a cutting, proceeds nearly parallel with the road to Leicester to its termination on a piece of land belonging to the Vicar of St. Mary's, at the end of Belgrave Gate, and nearly opposite to the Public Wharf. There appears to be no obstacle to the Railway being continued across Humberstone Gate to the London Road, and if this be not accomplished by its extension to form a junction with the London and Birmingham Railway, public convenience, and the interest of individuals in improving their property, will probably cause it to be effected.

Having thus completed a description of the line, I shall now advert to the Levels of the Railway. From Pinxton, for nearly two miles, the fall will be fourteen feet in each mile, and from thence to Long Eaton twelve feet in the mile. – Over the Valley of the Trent, for nearly two miles, the Railway will be level, and then rise uniformly four feet in each mile to its termination at Leicester.

In a distance of 34 Miles, it would be difficult to find a country more favourable for the formation of a Railway, with levels so well adapted to the contemplated traffic. In its whole course, there is no heavy or expensive work, – no extensive cutting or high embankment, but merely such a removal of earth as is required to raise the slight embankments in the meadow grounds, through which the Railway traverses, above the height of floods.

For the purpose of illustrating the effect, of moving force on this road, I shall state the ordinary work of one Horse upon the different levels, although steam power is intended to be the means of conveyance employed. On the level Railway, one Horse, at his ordinary speed, will convey 15 Tons; with the aid of 12 feet fall in a mile, he will draw 28 Tons, and ascending that inclination, 10 ⅓ Tons; but as the principal traffic will be *down* the portion of the line having this declivity, its fall becomes an advantage. Upon a road rising 4 feet in a mile, a Horse will convey 13 Tons, and the contrary way, 18 Tons. These weights are inclusive of the carriages. The effective work will, therefore, be less by about one fourth. The Liverpool and Manchester Railway being so well known by almost every one, I shall accompany this report with a comparative section of that line, and of the one now proposed, by which it will be seen that the line herein described, from the Trent to Leicester, is rather nearer to a level than the Liverpool Railway at Chat Moss, which is commonly supposed to be quite level.

In preparing the Estimate of the expense of the Midland Counties Railway, the object I have kept in view has been the construction of as perfect a Railway as, in the present state of knowledge, is attainable, as well with reference to the line and the levels, as to the strength, the durability and the permanent accuracy of the road.

For a double line of Railway, the width is calculated to be nine yards at the surface, allowing for the necessary slopes, and it is intended for the surface to be covered with a Bed, or foundation of Gravel, broken Stone or Furnace Cinders, of the thickness of eight inches, on which the Stone Blocks, to support the Rails, will be imbedded. The gauge of the Railway, or the space between the Rails, will be the same as that adopted on other public Railways, namely 4 Feet 8½ Inches, and the central space between the two lines, made usually equal to this, will be increased to six feet, a narrower interval having been found inconvenient.

The Rails will be formed of parallel Bars of wrought Iron, of a proper section, and in lengths of fifteen feet, supported at every three feet by Cast Iron Pedestals. The weight of the Rails is estimated at 40 lbs. to each single yard, and that of the pedestal with a cast iron wedge, 16 lbs. each. I prefer the parallel Bar to the elliptical or parabolic forms commonly used, in as much as greater strength is obtained by the Pedestal becoming a fulcrum, as well as a support, to resist the action of the load; and to obtain the same advantage at the joining of the Bars, they will be firmly united by means of larger and stronger Pedestals.

The Bridges and other works which, though forming part of, are not essential to the perfection, of a Railway, will be plain and simple, but substantial and sufficient for the object. The fences are estimated at the cost of common Post and Rails, three in height, with Quicksets. The following abstract of the Estimate of a double line of Railway is founded on the foregoing statement.

Excavations and embankments, including the forming of the whole line, ...	51,117
Diverting and straightening the rivers Erewash and Soar, ...	1,688
Bridges, Culverts, small Drains and Masonry, ...	16,248
Covering the Road with Gravel or small Stone and filling up the space about the Blocks, ...	11,900
Blocks, and laying down and fixing the Rails, ...	29,070
Rails, pedestals, keys, and pins ...	53,210
Alterations to Roads, to adapt them to the level of the Railway, ...	765
Fences and Gates therein, ...	8,500
Houses and Buildings removed,	600
Land, ...	20,400
Wharfs, Weighing Machines and Toll Houses	2,500
Contingencies, ...	19,002
Total expense,	£215,000

This Estimate, though ample for its object, appears to be so very considerably less than the cost of some other public Railways lately executed, that it is necessary I should advert to the cause of this difference, lest persons unacquainted with the subject should be led to infer, that the calculations upon which it is founded, are either erroneous, or that a less perfect Railway than I have described is intended to be made. To shew that such conclusion is not warranted, and to remove any

adverse impressions thence arising, I shall give a tabular view of the average cost of one Mile, of the Liverpool and Manchester Railway, and of the estimated cost of the London and Birmingham, of the Manchester and Sheffield, and of the Midland Counties Railways. The different expenses being stated under different heads, it will be at once seen how the great cost and excess of expenditure on those Railways arise, and that in works common to all, no greater variation exists in their amount, than advantageous position for obtaining materials will cause.

	LIVERPOOL & MANCHESTER	LONDON & BIRMINGHAM	MANCHESTER & SHEFFIELD	MIDLAND COUNTIES
Cuttings and embankments, forming the line of Railway ...	£8,133	£6,924	£2,380	£1,575
Bridges, Culverts and Masonry	3,965	3,116	1,000	550
Tunnels	1,122	2,225	3,000	—
Rails, Pedestals, Wedges & Pins	2,187	1,893	1,620	1,565
Blocks and Sleepers	662	915	667	720
Ballasting the road and laying down Rails	663	915	860	485
Fixed Machinery	200	—	232	—
Fences and Gates	425	675	454	250
Land	3,074	2,220	900	618
Contingencies ...	1,834	2,117	1,120	560
Average cost per Mile	£22,265	£21,000	£12,233	£6,323

From these comparative abstracts, it may be inferred, how much the cost of a Railway depends on the nature of the Country through which it has to pass. Where the line follows the natural course of extensive Vales, such as those of the Erewash and the Soar, little cost is required to adapt the ground to the proper levels, but when Hills and Vallies have to be intersected, and the different elevations of the Country have to be reduced, even to an approximation of uniformity of levels, this can only be accomplished at great labour and expense, by resorting to high embankments over Vallies, and deep cuttings and Tunnels through the Hills, which, besides their enormous cost, involve other large expenses in the construction of Bridges and Archways, for the convenient occupation of the land, severed by such works, and also further very heavy and continued charges, in preserving the proper levels of the Railway under the progressive settlement of the embankments, till time has rendered their consolidation perfect. The Slopes required in the excavations and embankments, also greatly add to the quantity of land to be purchased and to the compensations to be made for the inconvenience they occasion. The Land required for the London and Birmingham Railway will average Eleven and a half Acres per Mile, when Five Acres only will be required for the Midland Counties Railway, and the quantity of earth to be removed in the former, will be 107,000 yards on the average of a mile and in the latter only 30,000. An impartial consideration of these circumstances will explain the grounds on which the correctness of the Estimate is claimed to be relied on, and my experience in works of a similar nature enables me to say, with confidence, that every part of this Railway may be executed at the cost I have assigned.

It will be expected that I should advert to the probable amount of traffic and the Revenue to be derived therefrom, for the information of those disposed to become Subscribers, who are otherwise unaquainted with the facts on which such estimate is formed – although it is a subject on which opinion may and will differ, I shall endeavour to state impartially, in what way I conceive, that not only a liberal, but a large return may be expected on the Capital required to complete the undertaking. There already exist Canal communications from Pinxton to Leicester, and in the same Vallies through which the projected Railway will pass: It may therefore be supposed that the Canals will be formidable Rivals to the Railway, but of this, there is no fear, as the *whole* expense of conveyance by Railway, will not exceed the cost of *freight alone* upon the Canals; besides which the Railway will give dispatch and regularity the Canals are incapable of, as well as security against depredations so frequent from the opportunities offered by the slow speed of Canal conveyance, three days being required to effect a passage, which by Railway will be accomplished in as many hours, In addition to these disadvantages, the Trent and Soar are very subject to floods, which cause, not only detention, but risks so great to the narrow Boats, alone in use in the South, as to exclude a Coal trade to those parts, which otherwise would exist to a great extent; the length by the Canals is also 10 Miles more than by the proposed Railway.

A Rival has recently appeared to the Coal trade on these Canals, in the Leicester and Swannington Railway, which has been formed with a view of supplying Coal from the Liecestershire Coal Field, distant only Twelve to Eighteen Miles from Leicester, and in consequence of the greatly reduced cost of conveyance, it appears that a great portion of the Coal trade must eventually be lost to the Canals, as the expense of carrying Coal along these has been from 7s. to 8s. per ton, 2s.6d, being the cost of freight and the remainder, the rates charged by the Navigations. In anticipation however, of the new supply of Coal from the Leicestershire Coal Field, the rates have been lately reduced 1s.6d. per ton, still leaving the cost from 5s.6d. to 6s.6d. whilst their new rival is stated to effect this for only 2s.6d. Competition under such disadvantages could not exist, either with the Canals or the Coal Owners of Derbyshire and Nottinghamshire; who have, from time immemorial, furnished the Markets of Leicester and the Country South of Trent with Coal, and latterly to an extent of about 160,000 Tons annually; the production and sale of which, have given employment to a very numerous population, and support to others dependent on the prosperity of the district. The contemplation of the probable effect that would follow the loss of so extensive a trade, has led to the project of the Midland Counties Railway, not only for the purpose of preventing so disastrous a consequence, but with a view to extend the markets hitherto possessed, by the reduced prices at which Coal will reach them from the improved means and increased facilities of conveyance.

Although it must be expected that the new Leicestershire Coal Field will establish the introduction of its Coal, yet that district will be found to possess no advantage over this, which is not more than counterbalanced by other circumstances, and the great expansion of the trade that may be expected, will afford sufficient demand for both districts. It is certainly to be regretted, that the benefits, which arise from new discoveries and improvements cannot be obtained, without causing Individual loss; but the great Interests of Society require such protection. Inferiority is always the victim of Superiority – and as the Pack Horse has been superseded by good Roads and Carriages, so Canals will succomb to the mighty power of Steam and Railways.

Although the expenses of transit on the Canals from Pinxton to Leicester have been great, and the navigation of the Soar very defective and dangerous, yet the traffic conducted thereon, has enabled their Proprietors to make dividends exceeding Forty Thousand Pounds a year, and it is probable, that if the public had been made participators in the benefit of this extended trade, by a reasonable abatement of the Rates – new trade would have been attracted by it, and diverted from other channels, so as not greatly, if at all, to have reduced the income of the navigations. – No Competitor would, under these circumstances, have arisen, to deprive them of a revenue, which is said to have paid 200 per Cent on Soar Shares, 70 per Cent on those of the Erewash, and 18 per Cent on the Cromford and Leicester Canal Shares.

The Midland Counties Railway will offer to the public economy, security and dispatch, Rates reduced to One half of those taken by the Canals, and the cost of conveyance, One Third that of Canal freight only – coal will be delivered at Leicester, at prices from 8s. to 12s. per ton, instead of from 13s. to 17s. the late prices. With these reductions, there will remain for the Railway Proprietors, on the present extent of trade, an annual income of Twenty Thousand Pounds, and probably twice that amount will be derived from the new sources of traffic, which must obviously arise, when the conveyance of goods is rendered safe, cheap, and expeditious. A great extension of the Coal trade into new districts, will be consequent upon reduced charges. New markets may be expected in the neighbourhoods of Stamford, and Peterborough; also at Northampton, Bedford, and other places bordering on the Grand Junction Canal; and from the liberal disposition of the Grand Junction and Grand Union Canal Companies, already manifested in their desire to establish a London Coal trade from Derbyshire, for which the Soft Coals of this County are well suited and to which the Railway will afford great facilities, a new demand may be anticipated of an extent, limited only by the means of supply: the consumption of the Metropolis exceeding Two Million Tons of coals annually! Lime for agricultural purposes, has been excluded by the extravagant price consequent on the expenses of conveyance, although the quality of the land in Leicestershire requires the liberal use of this manure: – Before Agricultural produce was depressed, the Lime trade gave employment to four large establishments on the Cromford Canal for the supply of Lime to Boats; this trade is extinguished, although the local consumption for lime is unabated. Cheap conveyance will revive the demand, and provide a considerable revenue. The Mount Sorrel Granite, the best material for roads within the reach of the Midland Counties, will by its comparative small cost, be in large request for public Roads, and form another article of extensive trade. From these sources and from various others which it is impossible to anticipate, although certain where the passage of goods is rendered expeditious and cheap, it is by no means improbable that the Revenue may be increased to

Forty Thousand Pounds per annum.

The Railway will also be made subservient to the conveyance of Passengers and Parcels and probably the Mail, an object not inferior in importance to the conveyance of Merchandise, and perhaps equally productive of Income: Indeed, in the estimate of Income from the London and Birmingham Railway, all heavy goods are excluded, except such as travel by Fly Packet Boats, and in a line of 112 Miles, the annual Income from Passengers is estimated at £246,000, and from Parcels £23,000, including the cost of conveyance, and the calculation is made from the number of persons, who at present travel by public conveyances; if equal results arise, from the saving of *Time and Money* in the use of that Railway, that have followed the opening of the Liverpool and Manchester, the increase will be nearly threefold, as Mr. H. Booth stated in his evidence before the Lords' Committee on the London and Birmingham Railway Bill, that the number of passengers had increased between Liverpool and Manchester in the proportion of 450 to 1,200. – On the Sheffield and Manchester line, a distance of 43 miles, over a mountainous district, the income from the same sources is estimated at £24,000 a year. On the Liverpool and Manchester Railway of 31 Miles, the income from Passengers has yielded nearly £120,000 a year. It cannot be doubted that the completion of this line to Leicester, must lead to its junction with the London and Birmingham Railway, as this may be effected by an extension of only 18 Miles to Rugby, or by another and perhaps more preferable route of 33 miles to near Northampton. Northwardly, the country is very favourable for an extension by Derby, Belper, and Chesterfield to Sheffield; and with or without this latter extension, the Railway will become the great thoroughfare from Sheffield, Leeds, York, and the North of England, as well as Scotland, to the South and to the Metropolis. From the many natural obstacles the country on each side of the Railway here described presents, no other great Midland line from North to South can be made a rival to it, and whilst the Birmingham and London Railway can only be obtained at an expense exceeding this, by three times its cost on the average of a Mile, the extensive manufacturing interests of Leicester, Nottingham, Derby, Sheffield, and Leeds, and their districts, will give the Midland Counties Railway an importance equal to that of any other line in the kingdom. A permanent benefit will be secured to it by the very moderate rates required to ensure a remunerating Interest on its invested Capital, at once attracting trade by economy, and securing to itself protection from competition.

Although from what has been stated, it would appear that a revenue may be realised far exceeding the amount necessary to the success of the undertaking, yet the income arising from the existing trade, *without any increase* would afford a fair remuneration for the capital to be invested.

From the comparatively recent application of Steam Power to Loco-motion, and the limited experience that has hitherto been had, it may fairly be assumed that great improvements have yet to be made, before the system approaches the degree of perfection of which it is capable. In its present infant state, it is shown on the authority of those who have the best means of judging, amongst whom I will enumerate Messrs. James Walker, Tredgold, Nicholas Wood, Stephenson, and Rastrick, that the expense of Loco-motive Steam Power on a level Railway, is about one Farthing per Ton per Mile, and this is further confirmed by two years experience of William Crawshay, Esq. of Cyfarthfa Castle, South Wales, in the use of Loco-motive Engines on his Railways, and by similar experience of fifteen months on the Glasgow and Garnkirk Railway, particulars of which have been communicated by the Engineer, and will be found detailed at the end of this report. It appears, therefore, from these data, that the cost of Loco-motive power from the Derbyshire Collieries to Leicester, will be from 6d. to 8d. per Ton, the expense of Waggons being covered by the saving of charges at present incurred in wharfing, and stacking Coal, and loading it into Boats at the Collieries.

From the less perfect line of the Leicester and Swannington Railway, as is shown by the comparative sections accompanying this Report, the nearer position of the new Coal Field to Leicester will not be attended with an advantage corresponding with this inequality of distance, in the conveyance of its produce to that town; and as respects places Northward of Leicester, including the district of the Wreak River, its comparative disadvantages will be decidedly great. The small cost of the moving power when Loco-motive Steam Engines are employed, scarcely admits of a saving of which the public would be sensible, whilst the quality of Derbyshire Coal commands, by its superiority, from one to two Shillings per Ton more in price with consumers than the best Leicestershire Coal yet known, and from the great depth at which this coal is *expected* to be found, its existence in the New Field being not yet *proved,* it will be attended with expenses well understood by the practical coal

master, greatly exceeding the cost of producing Coal from Mines nearer to the surface, as is the case with the Collieries of Derbyshire and Nottinghamshire. The old Leicestershire Coal Field, at Coleorton, will find a better market for its produce on the Ashby Canal, and therefore its owner will not enter into a ruinous competition with those who possess decided advantages. In addition to the foregoing, it may be noticed that the small capital required for the Midland Counties Railway, which, as a single line, will be completed at a cost, not exceeding that of the single line now constructed between Leicester and Swannington, enables its proprietors to take as low a gross tonnage as will be required on the latter for its much shorter length. The expense of transit, therefore, will, at least, be equal on each.

The country connected with the projected Railway, is very favourable for the extensions contemplated to Derby and Nottingham, which are intended to form part of the general plan, and to be proceeded with in the year succeeding the passage of the Act for the first portion, between Pinxton and Leicester. The Towns of Nottingham and Derby will unite with the line last mentioned at Long Eaton, and not only form a communication with Leicester, in distance but little exceeding the Turnpike Roads, but also with each other as short as by the present Road. The Towns of Nottingham and Derby will, by the completion of a Railway communication with the Metropolis, be placed within a distance equal to six hours time for Passengers, and within twelve hours for Merchandise.

It is needless to point out the utility of speedy communication with the Ports of shipment, or place of consumption. Those acquainted with the nature of trade, know it to be of vital importance; that the loss, in fact, of a day, is often the loss of a season in foreign orders; and when expedition is accompanied with cheapness, it must not only be of local, but national benefit. Landed proprietors are interested, in obtaining increased means of fertilising the soil, and what is equivalent to it, a diminished cost of conveying its produce to market. Manufacturers are benefited by the economy afforded in their production, and all classes participate in the diffusion of intelligence and knowledge cause by increased facilities of communication.

I am, Gentlemen, Your obedient Servant,
WILLIAM JESSOP,
Butterley Hall, Nov. 25th, 1832.

Joseph Glynn's Report on Loco-motive Engines, &c. [1]

BUTTERLEY, NOVEMBER 5, 1832.

Sir,

I beg leave to submit to you the following observations in compliance with a request, which I had the honor to receive from the Committee of the Midland Counties Railway, that I should prepare a detailed Report, on the means, particulars, cost, and advantages, of inland communication for the transit of goods and passengers, by loco-motive Steam Power on Railroads, having especial reference to the Midland Counties.

Since the introduction of wrought iron bars for the formation of Railways, the successive practical improvements in their manufacture and formation into suitable shapes, and the continual additions which have been made to their weight, as the price of iron was reduced, have combined in causing them to reach a degree of perfection, they were not supposed capable of attaining, only a few years ago. The rails used at Darlington, on the public Railroad, weighed 28 pounds per yard in length; on the Liverpool Railway, 35 pounds to the yard; on the Newcastle and Carlisle Railway, now in progress, 45 to 50 pounds to the yard, are intended to be used; and it is generally considered, that rails from 12 to 15 feet long, and not less than 40 pounds to the yard in length, supported on pedestals of cast iron of 12 to 14 pounds weight, placed three feet apart, into which the rails are firmly keyed, should be used for public Railroads. These should rest upon large and broadly based blocks of stone, bedded solid in good materials.

A Railway of the present day, so formed, presents two continuous parallel lines of iron, capable of sustaining the action of machinery, and affording every requisite security of transit upon it at any speed practically attainable.

The width of the Manchester Railway and its branches has determined that of future Railways, and although it is to be regretted that it has been made so narrow as only 4 feet 8½ inches between the rails, yet it has effectually settled the question as to the gauge of others. The width of the Railway obviously limits the diameter of wheels, five feet being the largest used, but for carrying coal or heavy goods, wheels of three feet in diameter seem to be used by common consent as the best size. These should be made of case-hardened metal, and their axles should run in brass bearings with oil; perhaps it would be well to have them steeled or case-hardened. All the carriages should have springs, they contribute much to the preservation of the Railway, by causing the wheels to take an equal bearing upon it, and they render the motion much more easy.

For conveying coal, your staple article of traffic, the gross weight of each carriage may be five tons, of which the carriage itself may be about three-tenths, or thirty hundred weight.

On a level Railway, with the carriages, wheels and axles in perfect order, it requires but a very trifling effort to move them forward; with only that degree of attention to be expected from ordinary work people, the moving power is not more than a two hundredth part of the weight propelled, that is to say, a force of traction of 56 pounds will draw a load of five tons.

It is evident, therefore, that a rise of one in two hundred doubles the resistance, and that a fall of like inclination will continue the carriages in motion by their own gravity, – that a descent of 12 feet in a mile, or 1 in 440, lessens the force required by about 45 per cent., and that a rise of four feet in a mile, or 1 in 1320, increases it by about 15 per cent.

Your coal trade has this advantage of 1/440 from Pinxton to the Trent, and thence to Leicester, it has the resistance of 1/1320 to encounter; but when the nature of your traffic is considered, it seems to me it would be difficult, in crossing any country, to find a line of Railway so well adapted for it, or so suitable for the advantageous employment of loco-motive engines.

On the Manchester Railway, from the top of the inclined plane at Liverpool, in a distance of about 30 miles, about seven miles only are level; the rest of the line rises and falls, with several inclinations, varying from 1 in 2640, to 1 in 96, or 55 feet in a mile, there being a mile and a half each of ascent and descent in the last mentioned slope; the well-known part over Chat Moss has an inclination of 1 in 1200 or four feet and four-tenths in a mile. Yet, with these disadvantages, passengers are conveyed upon it in an hour and a half, and merchandise in two hours, from

Liverpool to Manchester by the loco-motive engines.

To show what rapid improvements have been made in these machines, I may state, that on the 25th of April, 1829, the Directors offered a premium of £500 for the most improved loco-motive engine, conditionally, that if it should weigh six tons, it should be capable of drawing after it, in daily work, on a level plane, a train of carriages of the gross weight of 20 tons, including the tender and water tank, at the rate of ten miles an hour.

"The Rocket," which gained the premium, weighed 4¼ tons, her tender 3 tons, and drew a gross weight of goods and carriages of 30 tons, at the rate of thirteen miles an hour. This engine had a cylindrical boiler, with a number of small tubes passing directly through the water from end to end, serving as flues; and the same plan, with various improvements, has since been followed in all the other engines on that Railway. "The Novelty," a beautiful engine, made by Braithwaite and Co., on an entirely new construction, failed from the extreme lightness of its parts, in competing with the "Rocket;" an increase of strength in a subsequent attempt might have remedied this defect, and it is surprising that it should not have been done.

Mr. James Walker having published a very able Report on the comparative advantages of employing fixed and loco-motive engines as they then were, in which he inclines to favour the former, Messrs. Stephenson and Locke printed a reply, and gave a very satisfactory account of the results and performances of several engines on the Liverpool Railway, but subsequent to their adoption in an improved form.

These engines act by the adhesion of their wheels on the Railway, varying from one-fifteenth to one-twentieth of the engine's weight, so that under all circumstances, on a level plane, a loco-motive engine weighing 7½ tons may safely be reckoned to exert a force of 7½ cwts., and to draw after it a gross weight of 75 tons..

Much more than this has been done, in various trials, on the Manchester Railway, but it is my wish to state what may be done in every day work, and calculated upon for practical and useful purposes, and I am borne out in the opinion I have expressed by the daily performances of the loco-motive engines employed at the iron-works and collieries near Merthyr Tydvil in Wales.

The engines employed by William Crawshay, Esq., on the railway at Hirwain, have their boilers constructed on a plan exactly the reverse of those on the Manchester Railway, they are composed of a system of small tubes through which the water circulates, whilst the fire is applied externally. The engine last made weighs five tons and draws a load of 50 tons gross weight, on a level Railway 2½ miles in length; it carries its own water and fuel for the trip, and requires no tender.

I am aware that "The Rocket," weighing 4¼ tons, has drawn 46¼ tons, gross load, at 13 miles an hour, – that "The Planet," weighing six tons, drew a load of merchandize, with the carriages, weighing 80 tons, at the rate of 12½ miles an hour from Liverpool to Manchester; and that those heavy and powerful engines, "The Sampson," and "The Goliah," have performed extraordinary feats, one of them having taken a load of 151 tons to Manchester in 2½ hours, but I shall content myself with stating that each loco-motive engine, such as I have before described, will convey a load of goods equal to the usual cargo of a wide boat on the Canals.

To prevent the nuisance arising from smoke, coke is now employed for fuel, of which for every ton of gross weight conveyed one mile, "The Rocket" consumed 91-100ths of a pound, "The Novelty" 53-100ths, and the "Sampson" only 3-10ths of a pound; it has even been stated in some recent experiments on the Liverpool Railway, a quarter of a pound of coke has been found sufficient to convey one ton gross weight, a mile. But it may be safely reckoned that one pound of coke will, in fair average daily work, take a ton of goods a mile, exclusive of the carriages.

Mr. Nicholas Wood, in the last edition of his book on Railways says, "That so long as the expense of one loco-motive engine does not exceed that of six horses and their attendants, then goods can be conveyed with the same expenditure of motive power at fifteen miles an hour upon a Railway that they can be conveyed at two miles an hour upon a Canal". He further says, that "this is not the only benefit resulting from the application of steam power to Railways, viz, that goods are conveyed with the same expenditure of motive power at the rate of fifteen miles an hour on a Railway, as at the rate of two miles an hour on a Canal. If it be attempted to augment the velocity on a Canal to three miles an hour, then one loco-motive engine on a Railway, will in six hours and two-thirds, perform the work of seventeen horses on a Canal".

Messrs. Stephenson and Locke estimate the cost of each loco-motive engine at £600, and for every five engines in actual work they allow one spare or extra engine in making each engine at work cost £720.

For interest, depreciation, and repairs they allow £104, and for wages of the engine men, fuel, oil, hemp, and tallow, they reckon £220 12s.10d., each engine's proportionate expense of maintaining the watering stations amounts to £6 15s., making a total of £331 7s.10d. annual expenditure. For this they reckon that 486,234 tons nett weight will be conveyed one mile, or 16,207 tons 30 miles, making the cost of conveying a ton of goods one mile 164/1000 of a penny, or 4 9/10 pence for 30 miles.

Mr. George Stephenson, in his Report on the proposed railway from Sheffield to Manchester, dated September 20th, 1831, after the great experience he had had in the use of loco-motive engines of various powers, and well knowing their performance in daily practice, says "Between Manchester and Stockport, I propose that engines of about 20 horse power be used." Each engine of this power, would make twelve single trips of seven miles in a day. The average weight of passengers and goods which the engine would take at each trip would be 40 tons. This would be equivalent to 480 tons conveyed in one direction per day, and the cost including the expenses of engine power and attendants upon the coaches and waggons would amount to £4 per day".

This portion of the line Mr. Stephenson calls very favourable ground, having an extremely moderate ascent of 1 in 792, or about 6⅔ feet in a mile. The performance of the engines is equal to 3,360 tons conveyed on this rise, and the cost of conveyance is 286/1000 of a penny, or rather more than one farthing per ton per mile. Mr. Stephenson further says, "From Stockport to the foot of the first inclined plane, I would recommend that engines of about 40 horse power be used, as the line ascends from Stockport with an average inclination of about one in 155, or 34 feet in a mile. Since so large a proportion in weight, of the articles carried, will consist of lime descending from the summit, and as lime carriages will be drawn up in the opposite direction empty, (together with a load of merchandise), the average load of each engine may be estimated at 50 tons. An engine will make six single trips of 14½ miles each per day, which is equivalent to 300 tons conveyed in one direction, and the expense of engine power, including all the necessary attendants, will be about £4.10s. per day".

We find the performance of each engine on this part of the line equal to 4,350 tons conveyed one mile at an expense of 25-100ths of a penny, or one farthing per ton per mile, the apparent difference of expense on these two portions of the line being accounted for by the descending lime trade on the latter. It may be remarked, that the opinion given by Mr. Stephenson, is in a great measure applicable to your proposed line of Railway, although you have not the very great inclinations to contend with he here contemplates.

William Crawshay, Esq., in a letter to Sir C. Dance, dated February 23rd, 1832, already before the public, after observing that facts of past performances of any kind, are more satisfactory than anticipations of the future, proceeds to state that in the past twelve months between the 1st of January 1831, and the 1st of January, 1832, the loco-motive engine he bought of Mr. Gurney, weighing only 33 cwt, conveyed 42,300 tons of coal, ironstone, and iron, exclusive of the carriages in which they were drawn, a distance of 2½ miles upon his Railroad at Hirwain, in journeys of from 20 to 30 tons, as suited their convenience. That the expense was £112 9s., or less than one farthing per ton per mile for the goods conveyed; and that had there been nearly double the work to do, the engine would have done it with little or no increased expense, as she was invariably working idle, for the purpose of keeping the boiler full, about one half of her time. It has before been remarked that Mr. Crawshay's last made engine, is much heavier and more powerful than this, and draws a proportionate load.

Passengers must be offered, in travelling by Railways, either speed or cheapness, or both; I apprehend you will not state a less rate of travelling than that of the Liverpool Railway carriages, viz. 20 miles an hour. If a train of heavy goods be dispatched immediately after the railway coaches, at the rate of ten miles an hour, each engine may make two single trips per day, between the collieries and Leicester, conveying 50 tons of goods, or nett weight, the average distance being about 30 miles. With a double line of railway this may easily be done, we have seen that much more has been done elsewhere. The estimates of Mr. George Stephenson apply to the cost of conveying passengers and merchandise. The statements of Mr. Robert Stephenson and Mr. Joseph Locke, confirmed by the experience of Mr. Crawshay, show the expense of conveying coal. The enormous expenditure of the Liverpool Railway Company affords no criterion. Let one loco-motive steam engine conveying coal on the projected Railway from Pinxton to Leicester – taking the average distance from the collieries to Leicester to be 30 miles – travel at the rate of ten miles an hour, conveying 50 tons of coal, exclusive of carriages, or 75 tons gross weight, and suppose the weight of the loco-motive engine ready for work to be 7½ tons, the tractive force to be 7½ cwts. or 840 lbs., moving at the rate of 10 miles an hour, or 880 feet in a minute, the number of horses power will be 22.4, and adding 15 per cent. more for the rise from the Trent to Leicester (or 3.36), we have 26 horses nearly for the power of the engine required.

Fifty tons nett weight conveyed 30 miles at an expenditure of a pound of coke per ton per mile requires 1500 pounds, or say 15 cwt. The descent from Pinxton to the Trent more than balancing the rise from the Trent to Leicester. Two single trips per day will take 30 cwt. of coke, costing 18s. and men's wages, oil, hemp, and tallow, 6s. per day, for 312 working days, together amount to £374. 8s. the expense of working such an engine for a year. If each engine costs £600 and there be one spare or extra engine for every five in actual work, then the cost of each engine at work will be £720.

Interest on £720, at 5 per cent	£36.	0s.	0d
Depreciation, at 10 per cent	£72.	0s.	0d
Repairs	£48,	0s.	0d
Fuel, wages, oil, hemp and tallow	£374.	8s.	0d
Total annual charge	£530.	8s.	0d

Fifty tons per day, for 312 days, are equal to 15,600 tons a year conveyed 30 miles, or 468,000 tons conveyed one mile, at an expense of 8$\frac{16}{100}$ pence for 30 miles, or .272$^{(3\frac{4}{125})}$) of a penny, being little more than a farthing per ton per mile, supposing the trains to return empty from Leicester. But if any loading can be obtained as back carriage, it will cause a corresponding diminution of the expenses, and it is obvious that 25 per cent. less fuel on the two trips will be consumed when the train of waggons returns empty.

I have endeavoured to show what has been done by loco-motive engines, and what you may expect from them even in their present state: we may fairly anticipate that they will be still farther improved..

The advantages arising from the use of loco-motive steam power, on well constructed Rail-roads, as a means of transit, are so numerous and so great, that they are almost beyond estimation; the facility of transit invariably increases communication and intercourse.

The number of passengers who travelled on the Liverpool and Manchester line, during the half year ending December 31, 1831, was 256,321, or about 1424 persons daily, yielding an income to the Railway Company, of £58,348 10s. on passengers alone, besides a large quantity of merchandize, &c., whereas the stage coaches employed before the Railway was formed, could only convey 688 passengers if completely full.

The road between Stockton and Darlington, did not afford passengers sufficient to support a single stage coach, drawn by a pair of horses; but since the Railway was made, there are four carriages upon it, each carrying 24 passengers. The Darlington Railway was at first a single line, but the increasing trade demands a double one, which is forming out of the profits of the first; and shares originally £100 have lately been selling at £285. The town of Stockton has increased its trade, and is now an extensive shipping port for the surrounding district, whilst the diminished price of coals, has introduced manufactures and steam engines, and added to the comforts of all classes of the people.

When the four great towns of Nottingham, Derby, Loughborough, and Leicester, containing an aggregate population of upwards of 135,000, shall have been connected by such means of communication, and more especially if we look forward to an extension of the Railway from Leicester to Northampton, and its junction with the London and Birmingham line, which will pass within a few miles, the benefits arising to the midland counties may be more easily imagined than described. The population of the counties of Nottingham, Derby, Leicester, and Northampton, amounts to nearly 840,000. They will, virtually, be brought nearer the metropolis, they will have a cheap and ready access to the great mart for their produce, they will receive a better price for it, and they will receive in return a better and cheaper supply of all the comforts and conveniences of life. I am aware that this is a very imperfect outline of the whole matter, and but an indistinct Report on the questions you have put into my hands, which, indeed, might require a volume. I have endeavoured to make it as concise as possible, leaving many points untouched.

I have the honour to be, Sir,
Your most obedient servant,
JOSPEPH GLYNN.

To the Chairman of the Committee
of the Midland Counties Railway.

General Observations [1]

In laying the following observations before the public, the author of them wishes it to be understood that he has not entered into any detail, either of the traffic, estimates, or inclinations, of the various Lines of Railway, of which these *observations* treat, there being several objections to a work of this kind containing much detailed information. In the first place, it could not be correct in any one particular, – no Railways being executed with precisely the same inclinations or curves, as are shewn upon the parliamentary plans. Great alterations have hitherto been made in almost all cases in both respects, and it is quite impossible for any person, not intimately acquainted with the actual progress of such works, to give any correct information on the subject.

The public are also aware how necessary it is not to put too much faith in parliamentary estimates of expensive works, experience having taught us that very few works of magnitude have ever been completed for the original estimates. A thousand things occur in the execution of a large work which no man can anticipate by looking at a drawing on paper, and difficulties and obstructions always arise, which never could be anticipated. The fact is, that if all those great undertakings which have been executed, and which are now going on in this country, had been at the commencement estimated to the amount of what they have cost, and will cost, most of them never would have been undertaken at all, although they have since been found profitable undertakings. This is explained by their resources of traffic having been as much under estimated in Parliament as the expense of construction, and in many cases more so. Railways which are cheap to execute without much excavation, embankment, or tunnelling, may be accurately estimated for.

It appears to me that a general opinion only can be formed of the probable success or failure of most Railways in progress; and this general opinion may be formed pretty accurately by a comparison of various Lines of Railway, and by calculating the amount and description of population.

One great objection to detailed information is, that the public generally do not understand it, and it is tiresome and irksome to them to read it: their opinions must be formed from information of a general kind; and it is this description of information that I propose laying before the public; and from having paid a great deal of attention to the progress of the various Lines of Railway in the *Midland Counties* and North of England, and having a thorough knowledge of the country through which they pass, and of the various towns and places which they connect; being also well acquainted with the coal fields in Lancashire, Yorkshire, Derbyshire, and Staffordshire; and of the course of the various canals which intersect the country, I believe that I can lay before the public, in a general form, an opinion upon the probable success or failure of many of these undertakings, which may be useful to those persons desirous of embarking in Railways as investments, and varying from any information which can generally be obtained, either from the reports and statements of Railway Companies, or from magazines, the latter collecting information from the former, which is always partially drawn out, and confined to the narrow district of the country through which the Railway runs; entirely omitting a most important point, – namely, the probability of future undertakings interfering and competing with those they applaud. It will be my object to endeavour to place this in a clear light before the public, with the effect produced upon any Railway by a competing Line being carried into effect, as well as the probability of any such Lines being made.

When I say a Line of Railway is expensive, I mean that it will cost from £20,000 to £30,000 a mile; and, when cheap, I mean from £12,000 to £15,000. In forming my opinion upon competing Lines of Railway, and in giving a preference to one over another, I am guided by the following consideration: – Supposing two Railways to exist, (both equal in length, expense, and inclinations,) between two towns, but carried in different directions; and suppose that one of these Lines has sufficient local traffic to make it pay a certain per centage upon the outlay, and the other has no local traffic. I contend that the Line with the local traffic will be able to abstract from the other Line the whole of the traffic from the towns at the termini, – and for this reason, that the expense of carrying passengers upon a Railway is not increased in proportion to the number carried, – that is, that where a large traffic is carried upon a Railway, the proportion of the expense of repairs, engines, &c., is less, when compared with the revenue received, than when a small traffic is carried; and, therefore, that a Line with a large local traffic, although it may have no other advantages over its competitor, would be enabled to carry the traffic from one terminus to another, at a much less expense than a Line with no local traffic. The same reasoning extends to two competing Lines of Railway, one of which may have only one object in view, whereas another may not only embrace that object, but may have others entirely independent of the object for which the former was projected.

I have considered that the expense of travelling is of as great importance to the generality of persons as time is, and that a majority of those who travel would be satisfied to spend half an hour longer on the road in a distance of two hundred miles, if they could be carried for a few shillings less money. This, also, has reference to competing Lines. Local or intermediate population, also, appears to me an object of great consequence in estimating the value of a Railway, because however many Lines of Railway may be made to compete with the traffic from the termini of an existing Railway, it is very unlikely that any will ever be made to compete for the intermediate traffic, situated upon the Line of such Railway.

I have also taken into consideration the great advantages those Railways possess which run through a mineral district, or one calculated for manufactures. And also when any town can be cheaper and better supplied with coal by the opening of a Railway in progress, than it is at present. It appears to me to be quite a fallacy to suppose, that Railways are not calculated for the conveyance of heavy goods, as there are at present existing many Railways which have been constructed at a great expense, and which are used for little else but the carriage of minerals, the tolls being small.

I will instance the Stockton and Darlington Railway, where stationary engines occur. The Leicester and Swannington Railway is another instance. This Line has both self-acting and stationary engine planes, and is paying from 8 to 10 per cent. But it is said, Look at the Liverpool and Manchester Railway? which certainly does not reap that profit on the carriage of goods which was anticipated. It must, however, always be remembered, that this Railway has to contend with an exceedingly cheap and easy navigation, and it is only a short Railway, consequently the convenience in the quick transit of goods is not so much felt between those two towns, as it will be in towns situated further apart, and more difficult of access by the present means of conveyance, especially where the canals are either circuitous, or contain a great number of locks.

It must be well known that almost all Railways are subject to be injured by competing projects, which are almost every day springing up in all directions. I am, however, convinced, that a large majority of persons who have shares in undertakings, placed in the situation of Competing Lines, and persons who may have invested money in such undertakings with a fair prospect of an ample return for their capital, are ignorant of the fact, that the Railway, in which they may hold so many shares, is placed in the critical position of having a Competing Line to fight with for a great proportion of its traffic, and which will, in all probability, reduce the expected revenue upon their Line, at least one half.

There are many Railways which at present have a fine prospect of a plentiful harvest, and the public appear to think the prospect will continue. The shares rise in proportion to the amount of traffic they may be at present carrying; when, perhaps, in two or three years, projects which are now in progress will be ready for operation, by which a large portion of the traffic, at present carried upon such Lines, will be abstracted from them, and they must eventually settle down into nothing more than a fair investment for the money subscribed for carrying them into execution.

I believe that Railways, well selected, and, above all, well managed, when completed, are the very best investments for spare capital at the present day, and that they will progress beyond any bounds which have yet been contemplated, I have no doubt; at the same time I know, that the shares in some of those Railways now finished, and some in progress of execution, are at a premium far above their real value as investments, whilst others are far below what they ought to be.

It is my present intention, in a future series of this work, when I have made myself more accurately acquainted with the Railways in the southern portion of England, to make similar observations upon them. And also to lay before the public my ideas as to the position in which the various Canals will be placed, and the effect produced upon them by the opening of the different Railways now constructing.

In concluding these remarks, I would say, that the opinions upon the various Railways, given in the following observations, are founded entirely upon my own judgment, and from information which I have gained in travelling through the country on geological excursions, and when attending the Committees of both Houses of Parliament, and in procuring all detailed information in the shape of Reports, Estimates, Sections, &c., which could be of service to a person who made this a subject of study. And with these materials I have formed an opinion from detail, which I am now laying before the public in a general form. I am, however, perfectly prepared, if the facts and opinions I have stated are called in question or contradicted, to support these facts, from more detailed information which I have at hand.

PROGRESS OF RAILWAYS

Very few people are aware of the extent to which the Railway system has already progressed in this country. Like all improvements it had, for a time, to contend against the prejudices of our countrymen, and its progress was slow, and its supporters were few and feeble. But since the opening of the Liverpool and Manchester Railway, where the powers of the locomotive engine were so fully developed, and by the skill and science made so astonishingly apparent, Railways have progressed rapidly. A Railway communication is now complete between Lancashire and Birmingham; in another year, there will be a complete Railway communication between London and Lancaster. This chain of communication is composed of the London and Birmingham to Birmingham, the Grand Junction to Liverpool and Manchester, the North Union to Preston, and the Preston and Lancaster to Lancaster. There will then only remain about sixty miles to Carlisle, for which distance two rival schemes are in contemplation. In two years, I imagine, a Railway communication will be complete from London to Newcastle-upon-Tyne, at any rate to Darlington. This chain will consist of the London and Birmingham to Rugby, or Stone Bridge, the Midland Counties, or Derby and Birmingham to Derby, the North Midland to Leeds, the York and North Midland to York, and the Great North of England to Newcastle-on-Tyne or Darlington. In this chain of communcation, a distance of about 280 miles, no inclination will exceed 16 feet in a mile. A Railway communication

is also in progress from Bristol and Exeter to Birmingham. This chain consists of the Bristol and Exeter, the Bristol and Gloucester, and the Gloucester and Birmingham. When this is made complete, a communication by Railway will exist between Bristol and Lancaster, on the west side of the island, and between Bristol and Newcastle-on-Tyne, through the interior of the country. These are the Main Lines of Railway, at present in progress, running North and South.

There are several communications from London to the South Coast, – namely, the London and Dover, the London and Brighton, and the London and Southampton Railways. At some future time, a Railway may be made from London to York, through Cambridge and Lincoln, constituting a third Main Line for the East Coast. This country is, however, so thinly populated, destitute of minerals, and entirely of an agricultural character, that many years may elapse before it is carried into effect; indeed the whole of the country, east of the London and Birmingham Railway, presents a very uninviting prospect for Railway speculators and capitalists.

There are three communications in progress, running east and west across the island, one from Maryport in Cumberland, through Carlisle and Newcastle to Sunderland. One from Liverpool through Manchester, Rochdale, and Wakefield, to Kingston-upon-Hull, and one from Bristol to London, which will, in a few years, be carried to some of the ports on the South-East Coast.

I have no doubt but that Railways will in time approximate in number to our turnpike roads, as the inhabitants of small towns will plainly see, that unless they obtain a Railway communication, they will be left behind by their neighbours; and, if they cannot afford to have one with the unexceptionable inclinations of the Great Main Lines, they will satisfy themselves with inferior inclinations, and less expeditious travelling. When gas was first introduced into our large towns, it was considered too expensive a commodity for the less populous and less wealthy places, but it has now been introduced into almost every small country town, especially amongst the coal districts.

In the course of three years, 1500 miles of Railway will, in all probability, be executed in England alone; and, calculating the average cost, inclusive of engines, at £20,000 a mile, £30,000,000 will have been spent in carrying them into execution. It is an unprecedented advance in the improvement of intercourse, and it is a plain speaking fact of the estimation in which this mode of travelling is held by the nation, both as an investment and as an improvement upon the old system. England is not, however, the only country in which Railways have made such rapid strides. Scotland has followed her example; and, although she has not yet been successful in obtaining a Railway from the mouth of the Clyde to the mouth of the Forth; that is, from the town of Greenock, through Paisley, Glasgow, Edinburgh, to Dunbar; yet she has attempted it, and has succeeded in procuring a Bill from Greenock to Glasgow, and from Ayr to Glasgow; and will no doubt, in a few years, complete the chain of communication. This country has also been much engaged in endeavouring to obtain a communication with England, and various projects have been set on foot, and there is no doubt that one of them will, before many years elapse, be completed.

Ireland also, has been stirring; and, although little has yet been done in that unfortunate country, yet a good deal has been attempted; and probably, in a short time, Government may find some project which they may think it worth their while to encourage and support. A Bill has been obtained for a Railway from Dublin to Drogheda, which will form a portion of the communication from Dublin to Belfast. A Bill has also passed the Legislature for a Railway from Belfast to Armagh, a portion of which is at present in course of execution.

America has probably outstripped even England, in her activity in forming Railways; indeed that nation appears to have considered it a point of great national importance, and Railways are spreading through every district of the States. Belgium, France, Holland, Germany, Prussia, and Russia, are all alive to the great importance of this mode of intercourse, and are preparing to follow the example of Britain and America.

LIVERPOOL, Feb. 10, 1838. Printed by J.F. Cannell, 50 Castle-Street.

BIRMINGHAM AND DERBY RAILWAY

Bill Obtained 1836.

The Bill for this Railway passed through Parliament without any opposition. The expense of carrying the works of this Railway into execution will be under the average cost of railways, from the very favourable nature of the country through which it passes. The inclinations of the line are extremely favourable; it may consequently be worked at a moderate expense when compared with most railways.

Commencing at Derby, it runs in almost a direct line to Burton-upon-Trent, and Alrewas, through Tamworth, Fazely, near Coleshill, and joins the London and Birmingham Railway about three miles from Birmingham. It opens out a communication from Leeds, Sheffield, Derby and York, as well as the North-east of England and Scotland, to Birmingham, Bristol, Gloucester, Worcester, Cheltenham, and the South-west of England, and, with the Branch to Stone Bridge, forms one of the Lines from Leeds, Edinburgh, &c., to London.

It does not possess much local population, the intermediate Towns on the Line not containing more than 20,000 inhabitants; but in consequence of its forming the only communication from the North-east to the South-west, it becomes an important Main Line of railway, and from its cheapness of execution is most likely to turn out a very profitable undertaking. By making the Branch to Stone Bridge, which place is situated on the London and Birmingham Railway, about ten miles south of Birmingham, the Company propose carrying a portion of the traffic from the North to London, and competing with the Midland Counties Railway for that traffic. I will state my reasons why I think they can do this successfully.

The Midland Counties Railway from Derby to Rugby is about 48 miles in length. The Birmingham and Derby (with the London and Birmingham) from Derby to the same place, is 57 miles, giving an advantage to the Midland Counties of 9 miles in distance. The Birmingham and Derby have to make only 8 miles of railway to enable them to carry this traffic, and the Midland Counties have to make 40 miles. The former has an easy country to pass through, as regards its inclinations and the expense of execution, whereas the Midland Counties Railway, from Leicester to Rugby, presents a very heavy section, and much rising ground, the excavations and embankments being very great.

The Birmingham and Derby has two objects in view: its main object is the traffic from the North to Birmingham and the South-west, and its other object is the traffic to London. It is, however, independent of the London traffic, but if the shareholders can obtain a portion of it, so much the better for them. The Birmingham and Derby will have the assistance of the London and Birmingham Railway in competing with the Midland Counties, as the London and Birmingham Company must naturally wish to carry any traffic they possess as great a distance upon their own Railway as possible, and as the Birmingham and Derby affords them the advantage of doing this for a distance of 17 miles more than the Midland Counties, they will consequently prefer that route. The North Midland Railway Company being an uninterested party, will send their London traffic on that Line which can convey it at the cheapest rate.

The Birmingham and Derby, in consequence of having a certain traffic to Birmingham, and which cannot be taken away from it, will be enabled to carry the London traffic at a much cheaper rate than if they were dependent upon it. The only object of the Midland Counties Railway being the London traffic to and from the North, and its local traffic, if you deprive it of the London traffic you destroy the Midland Counties Railway as an investment. Thus, then, the case stands: the Midland Counties has the advantage of 9 miles in distance, or, in consequence of its inferiority of inclinations, about fifteen minutes in time between Leeds and London. The Birmingham and Derby has the advantage in the cheapness of the Line, the superiority of its inclinations and summits, its entire independence of the London traffic, and the assistance it will obtain from the London and Birmingham Railway Company in promoting that route. I think I may, therefore, fairly conclude that it will at least obtain half of this traffic, and must, of course, seriously injure the Midland Counties Railway.

The Birmingham and Derby Railway is extremely well situated in another respect, for if ever a Line should be made from Manchester, either down the Trent or the Churnet Valley to London, and the East; this Railway will have a large increase of traffic arising from such Line. This line from Manchester was contended for, the last session of Parliament, and the Birmingham and Derby Railway Company projected a Line from Tamworth to Rugby, in order to accommodate the traffic from Manchester to London. The Tamworth and Rugby was 26 miles in length, and pursued its course from Tamworth, through Polesworth, Atherstone, and Nuneaton, to Rugby; it reduced the distance from Manchester and Leeds to London, as compared with the Stone Bridge Line, about 7 miles. It was however, thrown out in the Standing Orders' Committee, for a pretended omission of a very trifling nature. It had powerful opponents in the London and Birmingham, the Grand Junction, and Midland Counties Railway Companies, as well as land-owners. It appears to me that it must be again revived at some not very distant period. It may be said that if the Tamworth and Rugby should be made, the Stone Bridge Line will have paid for itself, as it is only eight miles long, and is exceedingly cheap in its execution. But even should this not be the case, it will always form the best communication from the North, – from Leeds, Derby, and Matlock, to Coventry, and the well-known and much-frequented towns of Leamington, Warwick, and Kenilworth. However the case may be, as it is only eight miles long, and very cheap, it is not of very great consequence.

One of the great features of the Birmingham and Derby Railway is the probability of its being used for supplying the Birmingham market, and the London and Birmingham Railway with coal and coke, from the Derbyshire collieries, as it so happens that that coal is peculiarly adapted to the purposes for which the coal which is used in the Birmingham manufactories is applied, namely, for smiths' furnaces. This coal is also found to make excellent coke for the steel furnaces at Sheffield; I should therefore suppose it is equally adapted for the locomotive engine; – if so, it must be of great advantage to the London and Birmingham and Grand Junction Railways, which are at present supplied at Birmingham in a great measure from the Lancashire collieries. Leamington, Warwick, and Kenilworth, and perhaps Coventry, will be partially supplied with this coal. If what I state should be the case, the coal trade alone will be sufficient to make the Line pay a fair per centage. With all these probabilities in its favour, I have become much attached to this Line, and I am convinced it is one of the best railway investments in England.

THE MIDLAND COUNTIES RAILWAY

Bill Obtained in 1836

The Bill for this Railway was opposed by a competing Line, the promoters of which were so far successful as to obtain a clause in the Bill, suspending the execution of that portion of their works from Leicester to Rugby for a year, to enable the competing Company to bring their scheme fairly before Parliament if they thought proper. They have not, however, done this; and the Midland Counties are now at liberty to execute the whole of their Line. They were also opposed

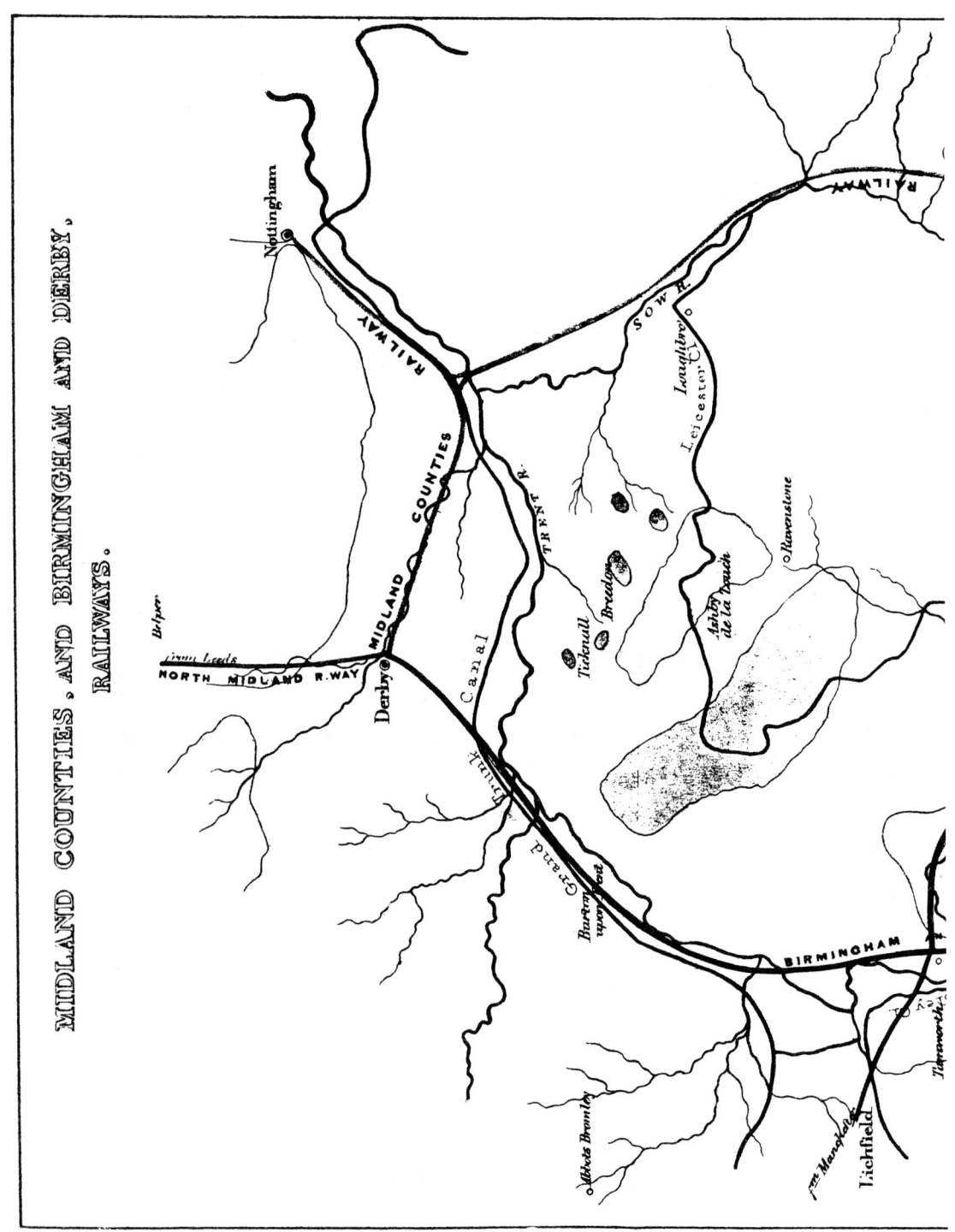

MIDLAND COUNTIES, AND BIRMINGHAM AND DERBY,
RAILWAYS.

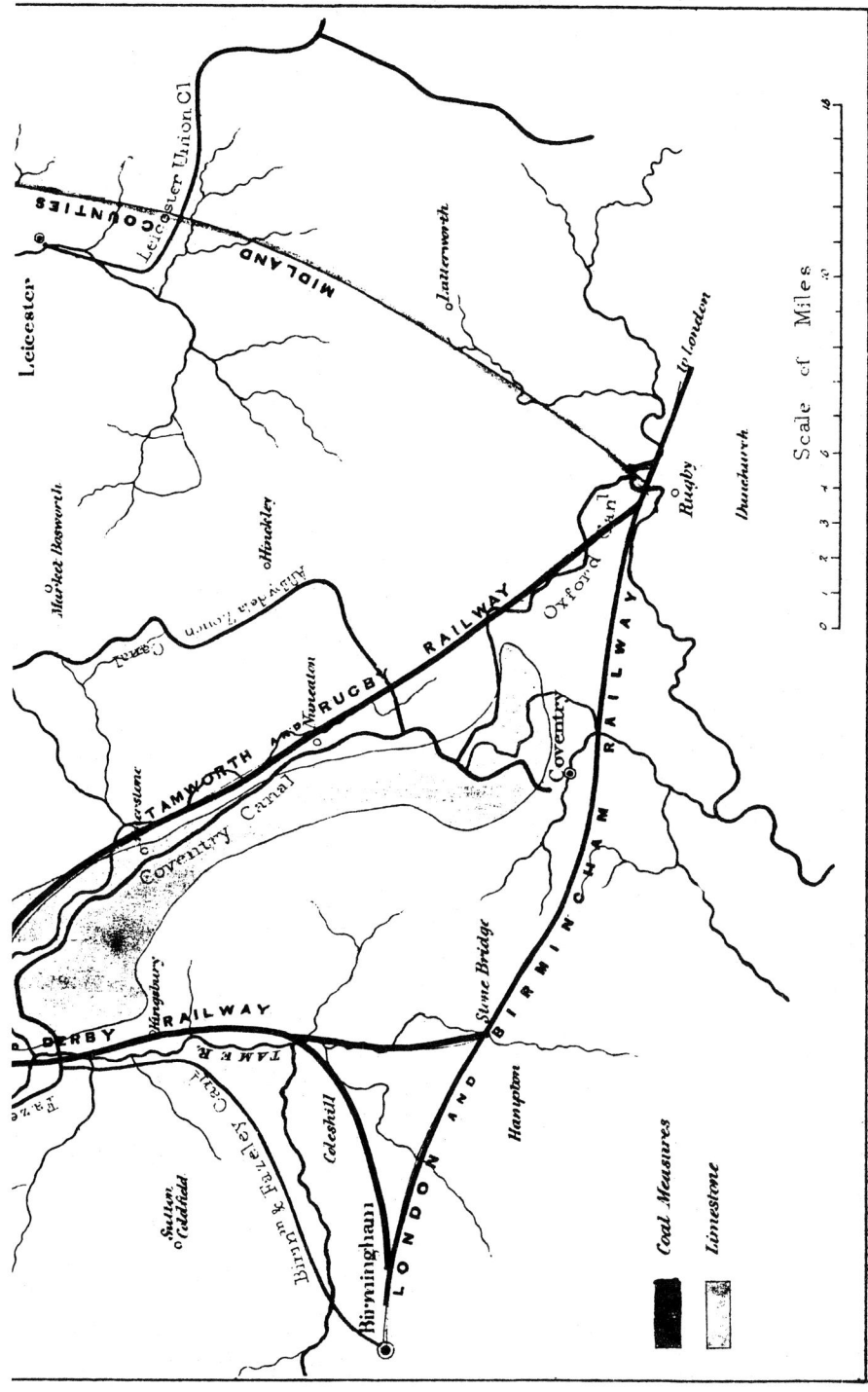

48. As well as the Midland Counties and Birmingham & Derby Junction Railways, the map accompanying J.F. Cannell's *General Observations* also indicated the proposed Tamworth and Rugby line. Editor's original.

by the coal proprietors of Leicestershire, and the Leicester and Swannington Railway Company, in consequence of their having carried the Line on the east side of the town of Leicester, which did not suit the convenience of these parties. They also appeared with a competing Line, or a deviation Line, which was, in my opinion, in every respect superior to the Line of the Midland Counties Railway; and why they had not adopted it in the first instance I cannot tell, unless it was an intentional omission, which is not unlikely, seeing that many of the most active of the Midland Counties Railway supporters are also Derbyshire coal owners, and it is well known these parties are interested in throwing any obstacle in the way of the Leicestershire coal owners, which could give the former an advantage in the trade.

This Railway appears beset with difficulties. On the west side it has the Birmingham and Derby, with its favourable levels and works to contend with, as a competing Line from the North to the South. On the east side it has another competing Line, which, although at present dormant, may again be revived. This Line commences near Leicester, and running through Market Harborough and Northampton, joins the London and Birmingham Railway at Blisworth, and although it is in every respect a bad Line, the inclinations being abrupt, and the works enormous, with little local population, yet it serves to harass and perplex the Midland Counties Railway.

The Midland Counties was projected for the purpose of carrying the whole of the traffic from Leeds, Sheffield, and the North-east of England to London, and on having that traffic the prosperity of the Line from the Trent to Rugby almost entirely depends. I have stated in my remarks on the Birmingham and Derby Railway why I think that Line, although 9 miles further round, will have other advantages over the Midland Counties to compensate for the loss of distance, and that it will at any rate have a fair share of the North and South traffic. And if the Tamworth and Rugby Railway be ever made, it will undoubtedly carry the whole of the North and South traffic, as it is as near to London, is a better Line in every respect, and having the Manchester traffic entirely at command, could well afford to carry that traffic on which the Midland Counties depends, at a much cheaper rate than they can do.

If the Midland Counties were free from all these difficulties I should say it was a fair investment, although a portion of it is very expensive, and its local population, considering the length of the

Line, is not great. But with these various difficulties surrounding it, I cannot help thinking that its ultimate success is very doubtful.

If the Company would content themselves with making that portion between Derby and Nottingham for the present, I think it would be a judicious conclusion, and let them wait a few years to see what progress railways make, as a very few years will decide whether all those railways which now stand as doubtful ones will repay the expense of construction. If they were to pursue this course, the only benefit withheld from the public would be the exclusion of the town of Leicester from a railway communication; which, although a flourishing town, is not by any means sufficient to support an undertaking of such magnitude as the Midland Counties Railway.

The Birmingham and Derby is placed in a very different position. It has, as I stated, a certain traffic that no railway can take from it, and by making a short line of 8 miles, it is opened out to compete for the whole of the Midland Counties Railway traffic; so that in the one case an experiment is tried on 8 miles of easy and cheap railway, and in the other case, the same experiment will have to be tried on 40 miles of expensive railway. By making the line from Derby to Nottingham, they would ensure a safe and speedy return for their capital, as that portion is easy and cheap to execute, the levels are very favourable, and the traffic is great.

A few years will either entirely set aside the Tamworth and Rugby, and the Northampton Line, or, by their being revived and made, will prove the prudence of the course I suggest; as, even if the Midland Counties be made, it will not in the least degree do away with the necessity of either of these Lines, if that necessity should be found to exist. I have heard of a proposition having been made either by one party or the other, to amalgamate the Birmingham and Derby and the Midland Counties Railways; but it appears to me that the interests of the two Companies are so diametrically opposed to each other, looking at them in any point of view, that I cannot see how this arrangement could be attended with any good result.

There was also another Line projected by the Midland Counties Railway Company, which, if made, might in some measure have competed with the North Midland for 20 miles: it was called the Pinxton Line: it ran from between Nottingham and Derby to near Chesterfield: it was an expensive Line, with tunnels, and has been abandoned.[2]

LEICESTER AND SWANNINGTON RAILWAY

This Railway has now been at work a few years. It extends from Leicester to Swannington, with a subsequent extension of a few miles. It is a single line of Railway, and is mostly used for the conveyance of coals, which are principally worked in the neighbourhood of Whitwick and Snibston, and carried along this Railway to Leicester; the greater quantity are consumed in Leicester, although a considerable quantity are sent by canal to Northampton, Leighton, Buzzard, and other towns in the South. Previous to the opening of this Railway, the town of Leicester was supplied with coal from Derbyshire, and the effect of the Leicestershire coal from Swannington being introduced into that market was a reduction of from about 18s. or 19s. a ton, to 10s. and 12s., thereby effecting a saving to the town of Leicester of from £40,000 to £50,000 a year.

This little Railway has, contrary to general anticipation, proved a good sound investment, and is now paying from 8 to 10 per cent per annum, and the shares are at a premium of about £25. The traffic in passengers is very small, and does not amount to more than £1000 per year. Slates, lime, granite, and coal, are almost the entire traffic, clearly proving that Railways are well calculated for the carriage of heavy merchandise. The trains travel at the rate of from 10 to 15 miles an hour, and heavy and very powerful locomotive engines are used. The inclinations on the line are abrupt, and there are both self-acting and stationary engine inclined planes, which are known to be very great drawbacks on all Railways, but more especially on those Lines used for the conveyance of passengers. The inclined plane at Bagworth is found a very great nuisance upon this Line, although the traffic is almost entirely confined to goods.

This Railway was for some time competed with by the Loughborough Canal Company, which canal conveyed the coals from the Derbyshire collieries to Leicester, previous to the opening of the Leicester and Swannington Railway, at which time the shares of the canal were valued and sold at nearly £5000 each; since the Railway has been in operation, they have fallen to about £1200 and £1500 a share, and I am inclined to believe that a great many canals must share the same fate.

I have been informed that some persons are endeavouring to find coals in the neighbourhood of Leicester; it is, however, exceedingly improbable that any coal exists there, and it is not unlikely that before long they may find themselves working in rather a hard material, as the granite does not lie a very long way from them. If, however, they should succeed in procuring any coal, it must be so dislocated from the eruptions which have taken place in the neighbourhood since it was deposited, that it cannot be worth working.

Reference Notes

NOTES TO PREFACE

1. Other than, that is, the contemporary guide books: R. Allen, *The Nottingham and Derby Railway Companion,* 1839 (republished by the Derbyshire Record Society, 1979); J. Drake, *Drake's Road Book of the Nottingham & Derby and Derby Junction Railways,* 1839; E. Allen, *The Midland Counties Railway Companion,* 1840; and R. Tebbutt, *A Guide or Companion to the Midland Counties Railway,* 1840. Other details have been included in the introductory chapters to histories of the Midland Railway: F.S. Williams, *The Midland Railway, its Rise and Progress,* first published 1876 and other editions to 1888 (5th and last republished by David & Charles, Newton Abbot, 1968); C.E. Stretton, *The History of the Midland Railway,* 1901; C.Hamilton Ellis, *The Midland Railway,* Ian Allan Ltd, 1953; E.G. Barnes, *The Rise of the Midland Railway, 1844–1874,* Allen & Unwin, 1966; and R. Williams, *The Midland Railway, a New History,* David & Charles, 1988.

NOTES TO INTRODUCTION

1. R.S. Smith, "Huntingdon Beaumont; Adventurer in Coalmines", *Renaissance and Modern Studies,* Sisson & Parker Ltd for the University of Nottingham, Vol.I, 1957, pp.115–153.
2. R.S. Smith, "England's First Rails: A Reconsideration", *Renaissance and Modern Studies,* Sisson & Parker Ltd for the University of Nottingham, Vol.IV, 1960, pp.119–134.
3. Northumberland Records Office, ZAN M 17/197 664. Information courtesy M.J.T. Lewis, and P. Stevenson, "Tramways and Railways in the Nottinghamshire Coalfield", *Journal of the Railway & Canal Historical Society,* Vol.XV No.3, July 1969, pp.45–54.
4. C. Hadfield, *The Canals of the East Midlands,* David & Charles, Newton Abbot, 1966, pp.36–41, 64–67, and 82–85.

5. Philip Riden, *The Butterley Company, 1790–1830,* published by the author at Wingerworth, 1973, pp.1–6.
6. R.J. Abbott, "The Railways of the Leicester Navigation Company", *Transactions of the Leicestershire Archaeological Society,* Vol.XXXI, 1955, and Hadfield, *op. cit.,* pp.85–92.
7. J.A. Birks and P. Coxon, "The Mansfield & Pinxton Railway," *The Railway Magazine,* July and August 1949, pp.224–30.
8. W.W. Tomlinson, *The North Eastern Railway: Its Rise and Development,,* first published 1914 and reprinted by David & Charles, Newton Abbot, 1967, p.97.
9. Ibid, quoting *Gentleman's Magazine,* February, 1825, p.114.
10. Ibid, quoting *Durham Chronicle,* 2nd October, 1824.
11. Ibid, quoting *Tyne Mercury,* 4th January, 1825.

NOTES TO CHAPTER 1, BACKGROUND AND EARLY SCHEMES

1. *Derby Mercury,* 20 and 27 October 1824.
2. Ibid, 3, 10, and 17 November 1824.
3. Ibid, 17 and 24 November 1824.
4. Ibid, 17 and 14 November, and 1 December 1824.
5. J. Marshall, *The Cromford & High Peak Railway,* David & Charles, Newton Abbot, 1982, p.3.
6. Riden, *op. cit.,* pp.6–13.
7. Marshall, *op. cit.,* p.3.
8. P.J. Long and the Reverend W.V. Awdry, *The Birmingham & Gloucester Railway,* Alan Sutton Publishing Ltd, 1987, p.4.
9. K.H. Vignoles, *Charles Blacker Vignoles: Romantic Engineer,* Cambridge University Press, 1982, p.30.
10. Riden, *op. cit.,* p.15.
11. Marshall, *op. cit.,* p.6.
12. *Derby Mercury,* 13, 20 and 27 October 1824.
13. Ibid, 8 and 15 December 1824.
14. Marshall, *op. cit.,* p.6.
15. *Derby Mercury,* 5, 12, and 19 January 1825.
16. Marshall, *op. cit.,* pp.9–10.
17. Ibid, p.11.

18. *Derby Mercury,* 19 and 26 January 1825.
19. Riden, *op. cit.,* pp.25–26.
20. *Derby Mercury,* 19 and 26 January 1825.
21. Ibid, 2 and 9 February 1825.
22. Ibid, 16 February 1825.
23. Ibid, 29 December 1824.
24. Ibid, 12 January 1825.
25. Ibid, 9 and 16 February 1825.
26. Ibid, 9 March 1825.
27. Ibid, 9 November 1825.
28. Tomlinson, *op. cit.,* p.161, quoting *Tyne Mercury,* 29 August 1826.
29. Ibid, p.162, quoting *The Scotsman,* 13 December 1826.
30. *Derby Mercury,* 13 and 27 October 1830.
31. Ibid, 27 October and 3 November 1830.
32. Ibid, 17 November 1830.
33. Ibid, 15 December 1830.
34. Ibid, 22 December 1830.
35. Ibid, 29 December 1830.

NOTES TO CHAPTER 2, STRUGGLING FOR RECOGNITION

1. Various prices were given by different authorities according to their individual sympathies, Sandars in the *Derby and Chesterfield Reporter* of 20 September 1832 quoting 17s 6d and 8s for the Leicester prices of the two sources of coal. Those quoted in the text were given by another correspondent, *William B.* in the same issue.
2. C.R. Clinker, "The Leicester and Swannington Railway", *Transactions of the Leicestershire Archaeological Society,* Vol.XXX, 1954, pp.59–114. Since republished by Avon Anglia.
3. W. Jessop, "Report to the Subscribers to the Midland Counties Railway," Butterley Hall, 15 November 1832, printed by G. Coates, Alfreton, 1833, and published with the prospectus. See appendices for the text of these items.
4. *PRO, Minutes of the Provisional Committee,* 3 March 1834.
5. Two different dates were given for this meeting, 28 August in *The*

Derby Mercury on 5 September 1832, and 27 August in *The Derby and Chesterfield Reporter* for 6 September, the latter falling on a Monday.

6. *Report of the Superintending Committee* from Alfreton, 15 October 1832. See Appendix 1.

7. Williams, *op. cit.*, 1st edition, p.9.

8. Ibid, p.11, quoting the Minute Books of the Eastwood Coalmasters.

9. *Derby and Chesterfield Reporter*, 20 September 1832.

10. *Derby Mercury*, 26 September 1832, and *Derby and Chesterfield Reporter*, 27 September 1832.

11. Ibid, 17 and 18 October 1832 respectively.

12. *Derby Mercury*, 24 October 1832.

13. Nottinghamshire Record Office, Deposited Railway Plan DP R4 (ex A3).

14. *Derby Mercury*, 13 February 1833, *Derby and Chesterfield Reporter*, 14 February 1833 and *Derbyshire Courier*, 16 February 1833.

15. *Derby Mercury*, 14 August 1833 and *Derby and Chesterfield Reporter*, 15 August 1833.

16. V.A. Hatley, "Northampton Hoodwinked? How a main line of railway missed the town a second time," *Journal of Transport History*, Vol.VII, No.3, May 1966, pp.160–72.

17. Williams, *op. cit.*, p.11.

18. *Derby Mercury*, 9 October 1833 and *Derbyshire Courier*, 12 October 1833.

19. *Derby Mercury*, 6 November 1833, *Derby and Chesterfield Reporter*, 7 November 1833 and *Derbyshire Courier*, 9 November 1833.

20. *Derby Mercury*, 6 and 13 November 1833.

21. Ibid, 4 and 18 December 1833, *Derby and Chesterfield Reporter*, 5 December 1833 and *Derbyshire Courier*, 7 December 1833.

22. *Derby Mercury*, 13, 20, and 27 November 1833 and *Derby and Chesterfield Reporter*, 14, 21, and 28 November 1833.

23. Nottinghamshire Record Office, Deposited Railway Plan DP R5 (ex A4).

24. *Derby Mercury*, 4 December 1833, 1 and 8 January 1834, *Derby and Chesterfield Reporter*, 3 and 26 December 1833, and *Derbyshire Courier*, 7 and 21 December 1833.

25. *Derby Mercury*, 22 and 29 January, 9 April 1834 and *Derbyshire Courier*, 5 April 1834.

26. *Derby Mercury*, 29 January, 5 and 12 February 1834.

27. Minutes of the Provisional Committee, 3 March 1834.

28. *Derby Mercury*, 5 and 12 March 1834, *Derby and Chesterfield Reporter*, 6 March 1834, *Derbyshire Courier*, 18 March 1834 and Minutes of the Provisional Committee, 3 March 1834.

29. *Derby Mercury*, 19 March 1834, *Derby and Chesterfield Reporter*, 20 March 1834, and *Derbyshire Courier*, 15 and 22 March 1834.

30. *Derby Mercury*, 23 April 1834, *Derby and Chesterfield Reporter*, 24 April 1834 and *Derbyshire Courier*, 26 April 1834.

31. *Derby Mercury*, 23 July 1834, *Derby and Chesterfield Reporter*, 17 and 24 July 1834, and *Derbyshire Courier*, 19 and 26 July 1834.

32. *Derby Mercury*, 3 September 1834.

33. Ibid, 10 September 1834.

34. Ibid, 14 January 1835.

35. *Derby and Chesterfield Reporter*, 18 February 1836.

36. *Derby Mercury*, 25 February 1835, and *Derby and Chesterfield Reporter*, 26 February 1835.

37. *Derby Mercury*, 15 April 1835.

38. Ibid, 12 November 1834, *Derby and Chesterfield Reporter*, 20 and 27 November 1834, and *Derbyshire Courier*, 8, 15, and 22 November 1834.

39. Nottinghamshire Record Office, Deposited Railway Plan DP R6 (ex A4)

40. *Derby and Chesterfield Reporter*, 5 and 19 February 1835.

41. *Derby Mercury*, 12 August 1835, and *Derby and Chesterfield Reporter*, 6 and 13 August 1835.

42. *Derby Mercury*, 2 and 9 September 1835, and *Derby and Chesterfield Reporter*, 10 September 1835.

43. *Derby Mercury*, 30 September 1835.

44. Vignoles, *op.cit.*, p.69.

45. Engineers' Reports, 11 November 1835.

46. Nottinghamshire Record Office, Deposited Railway Plan DP R7 (ex A5)

47. *Derby and Chesterfield Reporter*, 25 February 1836.

48. *Derby Mercury*, 9 December 1835.

49. Ibid, 27 January 1836.

50. Ibid, 20 April 1836.

51. Nottinghamshire Record Office, Deed Deposit, SD 3/29.

52. *Derby Mercury*, 13 April 1836, and *Derby and Chesterfield Reporter*, 14 April 1836.

53. Ibid, 4 February 1836.

54. Ibid, 14 April 1836.

55. Ibid, 17 February 1836.

56. Hatley, *op. cit.* pp.161 and 163.

57. Common Hall Books, Leicester.

58. *Derby Mercury*, 9 March 1836.

59. *Derby & Chesterfield Reporter*, 3 March 1836.

60. Ibid, 24 March 1836.

61. *Derby Mercury*, 27 April and 11 May 1836, and *Derby and Chesterfield Reporter*, 28 April 1836.

62. *Derby Mercury*, 11 May 1836, and *Derby and Chesterfield Reporter*, 5 May 1836.

63. *Derby and Chesterfield Reporter*, 12 May 1836.

64. *Derby Mercury*, 11 May 1836.

65. Ibid, 1 June 1836.

66. Ibid, 8 June 1836.

NOTES TO CHAPTER 3, CONSTRUCTION AND OPENING

1. 6 *Wm.IV, cap.lxxviii*, 21 June 1836.

2. Nottinghamshire Record Office, Deposited Railway Plan DP R9 (ex A6).

3. *Derby & Chesterfield Reporter*, 18 August 1836 and, for Woodhouse, Francis Whishaw, *Railways of Great Britain and Ireland*, first published 1842 and reprinted by David & Charles, Newton Abbot, 1959, p.334.

4. Hatley, op. cit. p.168.

5. *Derby & Chesterfield Reporter*, 18 August 1836.

6. *Derby Mercury*, 22 June 1836.

7. Ibid, 27 July and 3 August 1836.

8. Ibid, 2 November 1836.

9. Nottinghamshire Record Office, Deposited Railway Plan DP R10 (ex A7).

10. *Derby Mercury*, 8 March 1837, and *Railway Times*, No.23, p.159.

11. 3 *& 4 Vic., cap.cxxx*, 10 August 1840.

12. *Derby Mercury*, 26 April 1837.

13. *Board Minutes, No.73, 13 November 1838.*

14. *Ibid, No.99, 2 April 1839.*

15. *Common Hall Books, Leicester, 22 November 1837.*

16. *Committee for the Works South of Trent, Minute No.199, 12 December 1837.*

17. *Derby Mercury*, 11 November 1837.

18. 1 *& 2 Vic., cap.lvii*, 4 July 1838.

19. 22 *& 23 Vic., cap.lv.*

20. Midland Counties Railway Contracts.

21. *Derby Mercury*, 20 September 1837.

22. Midland Counties Railway Contracts.

23. Allen's *Guide, p.44.*

24. George Dow, *Railway Heraldry*, p.39. Photographic evidence (not of sufficient quality for reproduction) exists of the co-existence of the old main lines and new goods lines bridges.

25. Midland Counties Railway Contracts.

26. Committee for the Works South of Trent, Minute No.273, 28 March 1838.
27. *Derby Mercury,* 29 March 1838.
28. Committee for the Works South of Trent, Various Minutes between 18 November and 11 December 1837.
29. Board Minutes, No.25, 31 January 1837.
30. Ibid, No.87, 3 January 1839.
31. Ibid, No.89, 3 January 1839, and Wishaw, *op. cit.,* p.330.
32. Committee for the Works South of Trent, Minutes Nos.220 and 245, 1 January and 28 February 1838.
33. Ibid, Nos.259 and 273, 14 and 23 March 1838.
34. Ibid, No.580, 19 June 1839.
35. *Derby Mercury,* 11 April 1838.
36. Ibid, 1 August 1828.
37. Committee for the Works North of Trent, Minutes Nos.176 and 196, 8 May and 5 June 1838.
38. *Derby Mercury,* 22 August 1838.
39. Ibid, 1 August 1838.
40. Ibid, 12 September 1838.
41. *Nottingham Review,* 12 October 1838.
42. Committee for the Works South of Trent, Minutes Nos.479 and 480, 25 January 1839.
43. Committee for the Works North of Trent, Minute No.37, 29 January 1839.
44. *Derby Mercury,* 27 February 1839.
45. Ibid, 23 January 1839.
46. Ibid, 13 Febvruary 1839.
47. Board Minutes, No.86, 3 January 1839.
48. Committee for the Works North of Trent, Minute No.429, 1 May 1839.
49. Minutes of the Board and of the Committee for the Works South of Trent.
50. *Derby Mercury,* 1 May 1839.
51. Committee for the Works North of Trent, Minute No.391, 31 March 1839.
52. *Derby Mercury,* 3 February 1836.
53. Clinker, *op. cit.,* p.20.
54. Committee for the Works North of Trent, Minute No.170, 21 April 1838.
55. *2 & 3 Vic., cap.liii,* sections 44 and 47, 1 July 1839.
56. Drake's Guide.
57. Allen's Guide.
58. Whishaw, *op. cit.,* p.331.
59. *Records of the Borough of Nottingham,* Vol.IX.
60. Allen's Guide, p.3.
61. Ibid, pp.6–7.
62. *Derby Mercury,* 5 June 1839.
63. Ibid
64. Ibid, 17 July 1839.
65. Ibid, 15 August 1839.
66. Engineers' Reports.

67. *Derby Mercury,* 29 May 1839.
68. P. Stevenson, "The Building of the Midland Railway's Erewash Valley Extension and Trent Station", *Journal of the Railway & Canal Historical Society,* Vol.XXIII, No.4, October 1967, pp.53–61.
69. C.R. Clinker, *Register of Closed Stations,* 1962.
70. *Railway Times,* and *Herepath.*
71. Committee for the Works South of Trent, Minute No.658, 11 September 1839.
72. Nottinghamshire Record Office, Deposited Railway Plan DP R10 (ex A7).
73. Committee for the Works South of Trent, Minutes Nos.1, 2, and 43, 3 November 1836, and 17 January 1837.
74. Ibid, Nos.612, 616, and 618, 17, 20, and 27 July 1839. Also a copy of the Contractor's Bond in the Leicestershire Record Office.
75. Allen's Guide, pp.73–74.
76. Stretton, *op. cit.,* p.40.
77. John Gough, "Leicester (London Road) Station", *The Adaptation of Change, Essays upon the History of Nineteenth-century Leicester and Leicestershire,* edited by Daniel Williams and published by Leicestershire Museums, Publication No.18, 1980, pp.93–113. An unattributed plan of the 1846 layout accompanied an article in *LMS Magazine,* Vol.VII, 1930, pp.115–18: "Some Important LMS Passenger Stations, No.15. – Leicester (London Road)".
78. Committee for the Works South of Trent, Minute No.718, 20 November 1839.
79. Board Minutes, No.213, 18 February 1840.
80. Ibid, No.207, 18 February 1840.
81. Ibid, 30 January 1840.
82. Ibid, No.236, 10 April 1840.
83. Committee of Management, Minute No.727.
84. Board Minutes, Nos.727 and 878.
85. Allen's Guide, p.22.
86. Nottinghamshire Record Office, Deposited Railway Plan DP R72 (ex B1), 29 November 1847.
87. Board Minutes, No.248.
88. Ibid, No.262.
89. Committee of Management, Minute No.880, 1 February 1842.
90. Committee for the Works South of Trent, Minute No. 786, 25 March 1840.
91. Ibid, Minute No.776, 6 March 1840.
92. Committee for the Works South of Trent, Minutes Nos. 828 and 829, 27 April 1840.

93. Board Minutes, No.246, 4 May 1840.
94. Ibid, No.251, 4 May 1840.
95. Advertisement in *The Derby Mercury,* 6 May 1840.
96. Board Minutes, No.269, 29 June 1840.
97. *Derby Mercury,* 1 July 1840.
98. Allen's Guide, p.99.
99. Committee for the Works South of Trent, Minute No.166, 1 November 1837.
100. Ibid, No.255, 14 March 1838.
101. Ibid, No.560, 22 May 1839.
102. Board Minutes, No.205, 18 February 1840.
103. Ditto.
104. Ibid, No.230, 10 April 1840.
105. Committee for the Works South of Trent, Minute No.699, 23 October 1839.
106. Ibid, Minute No.198, 20 December 1837.
107. Whishaw, *op. cit.,* pp.238 and 331. See also J. Simmons, "Rugby Junction", *Dugdale Society Occasional Papers,* No.10, 1969.
108. Allen's Guide, pp.97–100.
109. *Herepath,* 11 May 1844, p.544.
110. *3 & 4 Vic. cap.cxxx,* sections 2, 3, and 4, 10 August 1840, and *5 Vic. Ssn.2 cap.ii,* 22 April 1842.
111. Whishaw, *op. cit,* p.327.

NOTES TO CHAPTER 4, *OPERATING THE RAILWAY*

1. *Derby Mercury,* 29 May 1839, and *Nottingham Review,* 28 June 1839.
2. *Derby Mercury,* 6 May 1840.
3. Ibid, 29 May 1839.
4. Ibid, 17 July 1839.
5. Ibid, 6 May 1840.
6. Via the Sheffield & Rotherham Railway.
7. *Nottingham Review,* 13 September 1839.
8. *Derby Mercury,* 20 May 1840.
9. Allen's Guide, *Table of Fares* in Appendices.
10. *Railway Times,* 11 July 1840.
11. *Derby Mercury,* 19 August 1840.
12. Ibid, 26 August 1840.
13. Ibid, 14 October 1840.
14. Ibid, 1 April 1840.
15. Ditto.
16. Ibid, 26 August 1840.
17. Ditto.
18. Ibid, 2 September 1840.
19. Ditto.
20. Ibid, 31 March 1841.
21. Ibid, 12 December 1840, and *Railway Magazine,* Vol.III, Nos. 68 and 70, 1840.

22. Minutes of Special Meetings of the Directors, 25 November 1840.
23. *Derby Mercury,* 23 December 1840.
24. Minutes of Special Meetings of the Directors, No.8, 16 November 1840.
25. Ibid, Nos. 1–3, 5 and 6, 16, and 18 November 1840.
26. *Herepath,* 23 January 1841.
27. *Derby Mercury,* 13 October 1841.
28. Committee of Management, Minute No.854, 20 January 1842.
29. Ibid, Minute No.825, 4 January 1842.
30. *Nottingham Review,* 16 December 1842.
31. Ibid, 27 January 1843.
32. Board Minutes, Nos. 773 and 774, 9 May 1843.
33. Ibid, Nos. 829 and 833, 12 June 1843.
34. *Derby Mercury,* 4 September 1839 and 12 December 1842.
35. Ibid, 3 June 1840.
36. Board Minutes, No.405, 5 November 1841.
37. Ibid, Nos. 1060 and 1062, 27 November 1843.
38. J. Marshall, *A Biographical Dictionary of Railway Engineers.* and Committee of Management Minute No.907, 1 March 1842.
39. Board Minutes, No.989, 20 September 1843.
40. Committee of Management, Minute No.884, 1 February 1842.
41. *Herepath,* 16 January and 6 February 1841.
42. *Derby Mercury,* 6 April 1842.
43. Ibid, 23 February 1842.
44. Ibid, 23 November 1842.
45. Ibid, 14 December 1842.
46. Ibid, 25 January 1843.
47. Ibid, 22 February 1843.
48. Ibid, 22 March 1843.
49. Ibid, 2 September 1840.
50. Board Minutes, No.283, 30 July 1840.
51. C.R. Clinker, ''The Birmingham & Derby Junction Railway'', *Dugdale Society Occasional Papers,* No.11, 1956. p.14.
52. *Derby Mercury,* 30 September 1840.
53. Board Minutes, No.317, 1 October 1840.
54. *Derby Mercury,* 17 August 1842.
55. Birmingham & Derby Junction Railway Board Minutes, 16 September 1842.
56. Board Minutes,, Nos.663–6, 1842.
57. Ibid, No.679, 9 January 1843.
58. *Derby Mercury,* 8 February 1843.
59. Board Minutes, No.708, 11 April 1843.
60. *Derby Mercury,* 26 April 1843.
61. Board Minutes,, Nos. 754, 777, and 790, 24 April, 9 and 22 May 1843.

62. *Derby Mercury,* 9 August 1843, and Board Minutes, No.881, 24 July 1843.
63. *Derby Mercury,* 16 August 1843, and Board Minutes, No.744, 25 April 1843.
64. Board Minutes, No.725, 11 April 1843.
65. Ibid, No.1017, 4 October 1843.
66. *Derby Mercury,* 16 August 1843.
67. Board Minutes, No.848, 5 July 1843.
68. Ibid, No.915, 1 August 1843.
69. Ibid, No.1045, 20 October 1843.
70. Ibid, No.684, 7 January 1843.
71. Ibid, No.854, 5 July 1843.
72. Ibid, No.895, 24 July 1843.
73. Ibid, No.852, 5 July 1843.
74. Ibid, No.886, 24 July 1843.
75. Ibid, No.1307, 11 May 1844.
76. Ibid, No.915, 1 August 1843.
77. *5 Vic. Sn II, cap.ii,* 22 April 1842.
78. Committee of Management, Minute No.871, 1 February 1842.
79. Board Minutes, No.698, 13 March 1843.
80. Ibid, No.953, 29 August 1843.
81. *Derby Mercury,* 20 September 1843.
82. Board Minutes, Nos. 1116 and 1126, 17 January 1844.
83. Ibid, No.1140, 5 February 1844.
84. *Nottingham Review,* 18 June 1841.
85. *Derby Mercury,* 20 April 1842.
86. Board Minutes, No.815, 12 June 1843.
87. Ibid, No.820, 12 June 1843.
88. *Derby Mercury,* 9 August 1843.
89. Board Minutes, Nos. 233 and 234, 10 April 1840.
90. *Derby Mercury,* 6 October 1841
91. Board Minutes, No.855, 5 July 1843.
92. *Nottingham Review,* 26 March 1841.
93. Board Minutes, No.1095, 20 December 1843.
94. Ibid, No.1270, 29 February 1844.
95. Ibid, No.884, 24 July 1843.
96. Ibid, No.745, 25 April 1843.
97. Ibid, No.718, 11 April 1843.
98. Ibid, No.1270, 29 February 1844.
99. Ibid, No.1287, 11 March 1844.
100. Committee of Management, Minute No.876, 1 February 1842.
101. Derby Council Minutes, 11 November 1841
102. *Derby Mercury,* 24 April 1843.
103. John Pudney, *The Thomas Cook Story.*.
104. *Derby Mercury,* 8, 15, and 29 June 1842.
105. *Nottingham Review,* 17 June and 1 July 1842.
106. Ibid, 30 June 1843.
107. *Derby Mercury,* 17 August 1843.
108. Ibid, 24 May 1843, and Board Minutes, No.803, 22 May 1843.
109. *Derby Mercury,* 12 December 1843.

NOTES TO CHAPTER 5, *EXTENSIONS AND AMALGAMATION*

1. *Derby Mercury,* 14 October 1840.
2. Ibid, 16 December 1840.
3. Ibid, 27 January 1841.
4. Ibid, 3 February 1841.
5. Ibid, 17 Febuary 1841.
6. Ibid, 24 February 1841.
7. Board Minutes, No.1047, 24 October 1843.
8. *Derby Mercury,* 5 June 1844.
9. Ibid, 10 July, 14 August and 11 September 1844.
10. *The North Staffordshire Railway,* by ''Manifold''.
11. Board Minutes, Nos.1264–6, 20 February 1844.
12. Ibid, No.1268, 20 February 1844, *Derby Mercury,* 21 February 1844, and *Herepath,* 24 February 1844.
13. Board Minutes, No.1270, 29 February 1844.
14. Ibid, No.1271, 29 February 1844.
15. Ibid, No.1272, 29 February 1844.
16. *Derby Mercury,* 6 March 1844, and *Herepath,* 9 March 1844.
17. Ibid, 3 April 1844 and 30 March 1844, respectively.
18. Board Minutes, No.1300, 1 May 1844.
19. *Derby Mercury,* 10 November 1841.
20. Board Minutes, No.1091, 20 December 1843.
21. Ibid, Nos. 1063 and 1097, 27 November and 20 December 1843.
22. Ibid, No.1120, 17 January 1844.
23. *Herepath,* 1 June 1844.
24. Board Minutes, No.849, 5 July 1843.
25. Ibid, No.918, 1 August 1843.
26. Ibid, No.924, 5 August 1843.
27. Ibid, No.926, 8 August 1843.
28. Ibid, No.925, 8 August 1843.
29. Ibid, No.930, 10 August 1843.
30. *Derby Mercury,* 16 August 1843.
31. Board Minutes, No.932, 12 August 1843
32. Ibid, No.960, 29 August 1843.
33. Ibid, Nos.969–70, 6 September 1843.
34. *Derby Mercury,* 27 September 1843.
35. Board Minutes, No.1007, 21 September 1843.
36. Ibid, No.1022, 4 October 1843.
37. Ibid, No.1021, 4 October 1843.
38. Ibid, No.1063, 27 November 1843.
39. Ibid, No.1021, 4 October 1843.
40. *Derby Mercury,* 17 July 1844.
41. *Herepath,* 27 July 1844.
42. Board Minutes, No.1032, 24 October 1843.
43. Ibid, Nos .1038, 1070, 1121, and 1261–2, 24 October and 27 November 1843, 17 January, and 20 February 1844.

44. Ibid, Nos.1274-6, 29 February 1844.
45. Thomas Bailey, *History of the County of Nottingham*, p.437.
46. *Illustrated London News*, 9 and 16 December 1843, pp.378–80 and 392–98.
47. *Derby Mercury*, 8 November 1843
48. Board Minutes, No.1123, 17 January 1844.
49. Ibid, No.1125, 17 January 1844.
50. Ibid, No.1249, 5 February 1844.
51. *Derby Mercury*, 13 March 1844.
52. Ibid, 3 April 1844.
53. *Herepath*, 18 May 1844.

NOTES TO CHAPTER 6, *LOCOMOTIVES*

1. No pretence at a comprehensive survey of the rolling stock of the company is claimed. It was originally the intention of the tutor to enlist the assistance of P.C. Dewhurst, an acknowledged expert on Midland Railway locomotives, who had written a series of articles in *The Locomotive* during the early nineteen-forties and had published a list of the Midland Counties stock in *British Railway Magazine* in February 1953 (Hamilton Ellis, *op.cit.*, pp.9–11). There are lists at pp.vi-vii of J.B. Radford's introduction to the 1979 reprint of Allen's *Nottingham and Derby Railway Companion* and at pp.15-18 of the late Bertram Baxter's *British Locomotive Catalogue, 1825–1923*, ed. David Baxter 1982, which refers to articles in *The Journal of the Stephenson Locomotive Society*, Vol.22, 1946, pp.85, 122, and Vol.15, 1949, p.61, and to *Locomotive News and Railway Contractor*, Vol.12, 1922, pp.19, 52. The details given in the text are taken primarily from itemised contemporary sources, annotated as required with additional information from the sources mentioned above.
2. Committee for the Works North of Trent, Minute No.146, 13 March 1839.
3. Ibid., Minute No.147.
4. *Nottingham and Derby Railway Companion*, Introduction, pp.vi, vii, and x.
5. Barnes, *op.cit.*, pp.27–28.
6. *Nottingham and Derby Railway Companion*, Introduction, p.x.
7. Baxter, *op. cit.*, pp.15–18.
8. *Nottingham and Derby Railway Companion*, Introduction, p.xi.
9. Clinker, *The Birmingham & Derby Junction Railway*, p.26.
10. J.B. Radford, *Derby Works and Midland Locomotives*, Ian Allan, 1971, p.20.

NOTES TO CHAPTER 7, *EPILOGUE*

1. Nottinghamshire Record Office, Deposited Railway Plan DP R70 (ex A64), 30 November 1846.
2. 8 & 9 Vic. cap189, 4 August 1845.
3. 9 & 10 Vic, caps.163 and 155 respectively.
4. Nottinghamshire Record Office, Deposited Railway Plan DP R80 (ex B9), 30 November 1853.
5. Minutes of Proprietors' Meetings, 19 May 1848, and Bradshaw's *Shareholders' Guide, Railway Manual & Directory, 1856*, p.180.
6. Herepath's *Railway & Commercial Journal*, 9 October 1852, p.1104.

NOTES TO APPENDIX 1, *THE 1832 PROSPECTUS*

1. Listed by George Ottley in his *Bibliography of British Railway History* as item 6930, from the Library of the Institution of Civil Engineers. The full title of the pamphlet reads *Midland Counties Railway. Prospectus of the projected Railway from Pinxton to Leicester, with Reports on the estimated cost of that undertaking, and on the application of Loco-motive Steam Power to Railways generally. Alfreton: Printed by George Coates. 1833*. In addition to the prospectus, set out in Appendix 1, the pamphlet contained William Jessop's *Report on the Railway, &c.*, set out in Appendix 2, the substance of a Report received from the Engineer of the Glasgow & Garnkirk Railway, referred to in Chapter 6, Joseph Glynn's *Report on the Loco-motive Engines, &c.*, set out in Appendix 3, and an abstract from the Evidence, *On the Advantages of Rail Roads*, given on the London & Birmingham Railway Bill before a Committee of the House of Lords. The text as reprinted here is taken from a copy of this pamphlet in the possession of Leslie Hales, which has been bound together with a number of other pamphlets published around the same date.

NOTES TO APPENDIX 2, *WILLIAM JESSOP'S REPORT*

1. As note 1 to Appendix 1.

NOTES TO APPENDIX 3, *JOSEPH GLYNN'S REPORT ON LOCO-MOTIVE ENGINES, &c.*

1. As note 1 to Appendix 1.

NOTES TO APPENDIX 4, *GENERAL OBSERVATIONS*

1. *A few General Observations on the Principal Railways executed, in progress, & projected, in the Midland Counties & North of England, with the Author's opinion upon them as investments. Illustrated with Maps*. A pamphlet printed by J.F. Cannell, 50 Castle-Street, Liverpool, 1838, retailed in London by Longman, Orme, Brown, Green, and Longmans. The author's introductory remarks are reproduced together with his statements on the Birmingham & Derby, Midland Counties, and Leicester & Swannington Railways.
2. This last paragraph is taken from the section describing the North Midland Railway, not otherwise reproduced. The author confuses the Pinxton section of the Midland Counties Railway with the extension to Clay Cross.

Index

RAILWAY AND CANAL HISTORICAL SOCIETY

FOUNDED 1954   INCORPORATED 1967

The Railway and Canal Historical Society was founded in 1954. Its objects are to bring together all those seriously interested in the history of transport, with particular reference to railways, waterways, and all matters associated with them: to encourage historical research; and to promote a high standard of publication – in the Society's Journal or elsewhere. While railways and canals form the basis of the Society, it recognizes that a serious study of their history is often inseparable from a study of associated modes of transport, such as river navigations, roads, docks, coastal shipping, and ferries.

For many, the biggest benefit which membership brings is the contact with others having similar interests. In addition to local Groups, for members living in particular areas,

a recent extension has been the formation of groups catering for particular interests – road history, docks and coastal shipping, and tramroads. Members also have the advantage of regular issues of the Society's Journal, Bulletin, and Book Reviews, as well as access to the extensive Research Index and to the Research Fund.

Full details of membership will gladly be sent on application to the Membership Secretary: R J Taylor, 64 Grove Avenue, Hanwell, London W7 3ES.

A full list of the Society's publications is available on application to the Sales Officer: C Rhodes Thomas, 23 Beanfield Avenue, Coventry CV3 6NZ, from whom the publications themselves can also be obtained.